Tests and Worksheets

SAXON Math™

HOMESCHOOL

8/7

with **Prealgebra**

Stephen Hake

John Saxon

SAXON™
PUBLISHERS

Saxon Publishers gratefully acknowledges the contributions of the following individuals in the completion of this project:

Authors: Stephen Hake, John Saxon

Editorial: Chris Braun, Matt Maloney, Brooke Butner, Brian E. Rice

Editorial Support Services: Christopher Davey, Jay Allman, Shelley Turner, Jean Van Vleck, Darlene Terry

Production: Alicia Britt, Karen Hammond, Donna Jarrel, Brenda Lopez, Adriana Maxwell, Cristi D. Whiddon

Project Management: Angela Johnson, Becky Cavnar

Printed in the United States of America

ISBN: 978-1-59-141324-0

Manufacturing Code: 28 29 30 0982 20

4500809745

CONTENTS

Introduction

Saxon Math 8/7—Homeschool Tests and Worksheets contains Facts Practice Tests, Activity Sheets, tests, and recording forms. Brief descriptions of these components are provided below, and additional information can be found on the pages that introduce each section. Solutions to the Facts Practice Tests, Activity Sheets, and tests are located in the *Saxon Math 8/7—Homeschool Solutions Manual*. For a complete overview of the philosophy and implementation of Saxon Math™, please refer to the preface of the *Saxon Math 8/7—Homeschool* textbook.

About the Facts Practice Tests

Facts Practice Tests are an essential and integral part of Saxon Math™. Mastery of basic facts frees your student to focus on procedures and concepts rather than computation. Employing memory to recall frequently encountered facts permits students to bring higher-level thinking skills to bear when solving problems.

Facts Practice Tests should be administered as prescribed at the beginning of each lesson or test. Sufficient copies of the Facts Practice Tests for one student are supplied, in the order needed, with the corresponding lesson or test clearly indicated at the top of the page. Limit student work on these tests to five minutes or less. Your student should keep track of his or her times and scores and get progressively faster and more accurate as the course continues.

About the Activity Sheets

Selected lessons and investigations in the student textbook present content through activities. These activities often require the use of worksheets called Activity Sheets, which are provided in this workbook in the quantities needed by one student.

About the Tests

The tests are designed to reward your student and to provide you with diagnostic information. Every lesson in the student textbook culminates with a cumulative mixed practice, so the tests are cumulative as well. By allowing your student to display his or her skills, the tests build confidence and motivation for continued learning. The cumulative nature of Saxon tests also gives your student an incentive to master skills and concepts that might otherwise be learned for just one test.

All the tests needed for one student are provided in this workbook. The testing schedule is printed on the page immediately preceding the first test. Administering the tests according to the schedule is essential. Each test is written to follow a specific five-lesson interval in the textbook. Following the schedule allows your student to gain sufficient practice on new topics before being tested over them.

About the Recording Forms

The last section of this book contains five optional recording forms. Three of the forms provide an organized framework for your student to record his or her work on the daily lessons, Mixed Practices, and tests. Two of the forms help track and analyze your student's performance on his or her assignments. All five of the recording forms may be photocopied as needed.

Facts Practice Tests and Activity Sheets

This section contains the Facts Practice Tests and Activity Sheets, which are sequenced in the order of their use in *Saxon Math 8/7—Homeschool*. Sufficient copies for one student are provided.

Facts Practice Tests

Rapid and accurate recall of basic facts and skills dramatically increases students' mathematical abilities. To that end we have provided the Facts Practice Tests. Begin each lesson with the Facts Practice Test suggested in the Warm-Up, limiting the time to five minutes or less. Your student should work independently and rapidly during the Facts Practice Tests, trying to improve on previous performances in both speed and accuracy.

Each Facts Practice Test contains a line for your student to record his or her time. Timing the student is motivating. Striving to improve speed helps students automate skills and offers the additional benefit of an up-tempo atmosphere to start the lesson. Time invested in practicing basic facts is repaid in your student's ability to work faster.

After each Facts Practice Test, quickly read aloud the answers from the *Saxon Math 8/7—Homeschool Solutions Manual* as your student checks his or her work. If your student made any errors or was unable to finish within the allotted time, he or she should correct the errors or complete the problems as part of the day's assignment. You might wish to have your student track Facts Practice scores and times on Recording Form A, which is found in this workbook.

On test day the student should be held accountable for mastering the content of recent Facts Practice Tests. Hence, each test identifies a Facts Practice Test to be taken on that day. Allow five minutes on test days for the student to complete the Facts Practice Test before beginning the cumulative test.

Activity Sheets

Activity Sheets are referenced in certain lessons and investigations of *Saxon Math 8/7—Homeschool*. Students should refer to the textbook for detailed instructions on using the Activity Sheets. The fraction manipulatives (on Activity Sheets 1–6) may be color-coded with colored pencils or markers before they are cut out.

64 Multiplication Facts
For use with Lesson 1

Name _____

Time _____

Multiply.

6 × 8	5 × 7	3 × 3	6 × 2	4 × 7	9 × 3	8 × 5	2 × 4
7 × 2	4 × 5	8 × 2	8 × 6	2 × 9	5 × 6	9 × 7	4 × 9
8 × 9	7 × 9	2 × 6	3 × 8	7 × 8	9 × 6	3 × 2	6 × 7
5 × 2	3 × 7	8 × 7	6 × 3	2 × 2	7 × 7	9 × 8	4 × 3
7 × 6	8 × 8	4 × 8	3 × 5	8 × 3	9 × 5	2 × 7	5 × 8
6 × 6	2 × 3	4 × 4	5 × 3	9 × 9	3 × 9	8 × 4	7 × 3
4 × 6	7 × 5	3 × 6	6 × 9	5 × 4	9 × 4	2 × 5	6 × 4
5 × 9	3 × 4	6 × 5	2 × 8	7 × 4	4 × 2	5 × 5	9 × 2

A | 64 Multiplication Facts
For use with Lesson 2

Name _____

Time _____

Multiply.

6 × 8	5 × 7	3 × 3	6 × 2	4 × 7	9 × 3	8 × 5	2 × 4
7 × 2	4 × 5	8 × 2	8 × 6	2 × 9	5 × 6	9 × 7	4 × 9
8 × 9	7 × 9	2 × 6	3 × 8	7 × 8	9 × 6	3 × 2	6 × 7
5 × 2	3 × 7	8 × 7	6 × 3	2 × 2	7 × 7	9 × 8	4 × 3
7 × 6	8 × 8	4 × 8	3 × 5	8 × 3	9 × 5	2 × 7	5 × 8
6 × 6	2 × 3	4 × 4	5 × 3	9 × 9	3 × 9	8 × 4	7 × 3
4 × 6	7 × 5	3 × 6	6 × 9	5 × 4	9 × 4	2 × 5	6 × 4
5 × 9	3 × 4	6 × 5	2 × 8	7 × 4	4 × 2	5 × 5	9 × 2

64 Multiplication Facts
For use with Lesson 3

Name _____

Time _____

Multiply.

6 × 8	5 × 7	3 × 3	6 × 2	4 × 7	9 × 3	8 × 5	2 × 4
7 × 2	4 × 5	8 × 2	8 × 6	2 × 9	5 × 6	9 × 7	4 × 9
8 × 9	7 × 9	2 × 6	3 × 8	7 × 8	9 × 6	3 × 2	6 × 7
5 × 2	3 × 7	8 × 7	6 × 3	2 × 2	7 × 7	9 × 8	4 × 3
7 × 6	8 × 8	4 × 8	3 × 5	8 × 3	9 × 5	2 × 7	5 × 8
6 × 6	2 × 3	4 × 4	5 × 3	9 × 9	3 × 9	8 × 4	7 × 3
4 × 6	7 × 5	3 × 6	6 × 9	5 × 4	9 × 4	2 × 5	6 × 4
5 × 9	3 × 4	6 × 5	2 × 8	7 × 4	4 × 2	5 × 5	9 × 2

A | 64 Multiplication Facts

For use with Lesson 4

Name _____

Time _____

Multiply.

6 × 8	5 × 7	3 × 3	6 × 2	4 × 7	9 × 3	8 × 5	2 × 4
7 × 2	4 × 5	8 × 2	8 × 6	2 × 9	5 × 6	9 × 7	4 × 9
8 × 9	7 × 9	2 × 6	3 × 8	7 × 8	9 × 6	3 × 2	6 × 7
5 × 2	3 × 7	8 × 7	6 × 3	2 × 2	7 × 7	9 × 8	4 × 3
7 × 6	8 × 8	4 × 8	3 × 5	8 × 3	9 × 5	2 × 7	5 × 8
6 × 6	2 × 3	4 × 4	5 × 3	9 × 9	3 × 9	8 × 4	7 × 3
4 × 6	7 × 5	3 × 6	6 × 9	5 × 4	9 × 4	2 × 5	6 × 4
5 × 9	3 × 4	6 × 5	2 × 8	7 × 4	4 × 2	5 × 5	9 × 2

64 Multiplication Facts

For use with Lesson 5

Name _____

Time _____

Multiply.

6 × 8	5 × 7	3 × 3	6 × 2	4 × 7	9 × 3	8 × 5	2 × 4
7 × 2	4 × 5	8 × 2	8 × 6	2 × 9	5 × 6	9 × 7	4 × 9
8 × 9	7 × 9	2 × 6	3 × 8	7 × 8	9 × 6	3 × 2	6 × 7
5 × 2	3 × 7	8 × 7	6 × 3	2 × 2	7 × 7	9 × 8	4 × 3
7 × 6	8 × 8	4 × 8	3 × 5	8 × 3	9 × 5	2 × 7	5 × 8
6 × 6	2 × 3	4 × 4	5 × 3	9 × 9	3 × 9	8 × 4	7 × 3
4 × 6	7 × 5	3 × 6	6 × 9	5 × 4	9 × 4	2 × 5	6 × 4
5 × 9	3 × 4	6 × 5	2 × 8	7 × 4	4 × 2	5 × 5	9 × 2

B 30 Equations
For use with Lesson 6

Name _____

Time _____

Find the value of each variable.

$a + 12 = 20$ $a =$	$b - 8 = 10$ $b =$	$5c = 40$ $c =$
$\dfrac{d}{4} = 12$ $d =$	$11 + e = 24$ $e =$	$25 - f = 10$ $f =$
$10g = 60$ $g =$	$\dfrac{24}{h} = 6$ $h =$	$17 = j + 8$ $j =$
$20 = k - 5$ $k =$	$30 = 6m$ $m =$	$9 = \dfrac{n}{3}$ $n =$
$18 = 6 + p$ $p =$	$5 = 15 - q$ $q =$	$36 = 4r$ $r =$
$2 = \dfrac{16}{s}$ $s =$	$5 + 7 + t = 20$ $t =$	$u - 15 = 30$ $u =$
$8v = 48$ $v =$	$\dfrac{w}{3} = 6$ $w =$	$21 - x = 12$ $x =$
$y + 8 = 12$ $y =$	$36 = 3z$ $z =$	$\dfrac{48}{a} = 4$ $a =$
$b - 12 = 15$ $b =$	$75 = 3c$ $c =$	$\dfrac{d}{12} = 6$ $d =$
$36 = f + 24$ $f =$	$g - 24 = 24$ $g =$	$12h = 12$ $h =$

B | 30 Equations
For use with Lesson 7

Name _____

Time _____

Find the value of each variable.

$a + 12 = 20$ $a =$	$b - 8 = 10$ $b =$	$5c = 40$ $c =$
$\dfrac{d}{4} = 12$ $d =$	$11 + e = 24$ $e =$	$25 - f = 10$ $f =$
$10g = 60$ $g =$	$\dfrac{24}{h} = 6$ $h =$	$17 = j + 8$ $j =$
$20 = k - 5$ $k =$	$30 = 6m$ $m =$	$9 = \dfrac{n}{3}$ $n =$
$18 = 6 + p$ $p =$	$5 = 15 - q$ $q =$	$36 = 4r$ $r =$
$2 = \dfrac{16}{s}$ $s =$	$5 + 7 + t = 20$ $t =$	$u - 15 = 30$ $u =$
$8v = 48$ $v =$	$\dfrac{w}{3} = 6$ $w =$	$21 - x = 12$ $x =$
$y + 8 = 12$ $y =$	$36 = 3z$ $z =$	$\dfrac{48}{a} = 4$ $a =$
$b - 12 = 15$ $b =$	$75 = 3c$ $c =$	$\dfrac{d}{12} = 6$ $d =$
$36 = f + 24$ $f =$	$g - 24 = 24$ $g =$	$12h = 12$ $h =$

A   **64 Multiplication Facts**
For use with Lesson 8

Name _____

Time _____

Multiply.

6 × 8	5 × 7	3 × 3	6 × 2	4 × 7	9 × 3	8 × 5	2 × 4
7 × 2	4 × 5	8 × 2	8 × 6	2 × 9	5 × 6	9 × 7	4 × 9
8 × 9	7 × 9	2 × 6	3 × 8	7 × 8	9 × 6	3 × 2	6 × 7
5 × 2	3 × 7	8 × 7	6 × 3	2 × 2	7 × 7	9 × 8	4 × 3
7 × 6	8 × 8	4 × 8	3 × 5	8 × 3	9 × 5	2 × 7	5 × 8
6 × 6	2 × 3	4 × 4	5 × 3	9 × 9	3 × 9	8 × 4	7 × 3
4 × 6	7 × 5	3 × 6	6 × 9	5 × 4	9 × 4	2 × 5	6 × 4
5 × 9	3 × 4	6 × 5	2 × 8	7 × 4	4 × 2	5 × 5	9 × 2

Saxon Math 8/7—Homeschool

B	**30 Equations**	Name _____
	For use with Lesson 9	Time _____

Find the value of each variable.

$a + 12 = 20$ $a =$	$b - 8 = 10$ $b =$	$5c = 40$ $c =$
$\dfrac{d}{4} = 12$ $d =$	$11 + e = 24$ $e =$	$25 - f = 10$ $f =$
$10g = 60$ $g =$	$\dfrac{24}{h} = 6$ $h =$	$17 = j + 8$ $j =$
$20 = k - 5$ $k =$	$30 = 6m$ $m =$	$9 = \dfrac{n}{3}$ $n =$
$18 = 6 + p$ $p =$	$5 = 15 - q$ $q =$	$36 = 4r$ $r =$
$2 = \dfrac{16}{s}$ $s =$	$5 + 7 + t = 20$ $t =$	$u - 15 = 30$ $u =$
$8v = 48$ $v =$	$\dfrac{w}{3} = 6$ $w =$	$21 - x = 12$ $x =$
$y + 8 = 12$ $y =$	$36 = 3z$ $z =$	$\dfrac{48}{a} = 4$ $a =$
$b - 12 = 15$ $b =$	$75 = 3c$ $c =$	$\dfrac{d}{12} = 6$ $d =$
$36 = f + 24$ $f =$	$g - 24 = 24$ $g =$	$12h = 12$ $h =$

| **A** | **64 Multiplication Facts**
For use with Lesson 10 | Name _____
Time _____ |

Multiply.

6 × 8	5 × 7	3 × 3	6 × 2	4 × 7	9 × 3	8 × 5	2 × 4
7 × 2	4 × 5	8 × 2	8 × 6	2 × 9	5 × 6	9 × 7	4 × 9
8 × 9	7 × 9	2 × 6	3 × 8	7 × 8	9 × 6	3 × 2	6 × 7
5 × 2	3 × 7	8 × 7	6 × 3	2 × 2	7 × 7	9 × 8	4 × 3
7 × 6	8 × 8	4 × 8	3 × 5	8 × 3	9 × 5	2 × 7	5 × 8
6 × 6	2 × 3	4 × 4	5 × 3	9 × 9	3 × 9	8 × 4	7 × 3
4 × 6	7 × 5	3 × 6	6 × 9	5 × 4	9 × 4	2 × 5	6 × 4
5 × 9	3 × 4	6 × 5	2 × 8	7 × 4	4 × 2	5 × 5	9 × 2

64 Multiplication Facts

For use with Test 1

Name _____

Time _____

Multiply.

6 × 8	5 × 7	3 × 3	6 × 2	4 × 7	9 × 3	8 × 5	2 × 4
7 × 2	4 × 5	8 × 2	8 × 6	2 × 9	5 × 6	9 × 7	4 × 9
8 × 9	7 × 9	2 × 6	3 × 8	7 × 8	9 × 6	3 × 2	6 × 7
5 × 2	3 × 7	8 × 7	6 × 3	2 × 2	7 × 7	9 × 8	4 × 3
7 × 6	8 × 8	4 × 8	3 × 5	8 × 3	9 × 5	2 × 7	5 × 8
6 × 6	2 × 3	4 × 4	5 × 3	9 × 9	3 × 9	8 × 4	7 × 3
4 × 6	7 × 5	3 × 6	6 × 9	5 × 4	9 × 4	2 × 5	6 × 4
5 × 9	3 × 4	6 × 5	2 × 8	7 × 4	4 × 2	5 × 5	9 × 2

1 | Halves

For use with Investigation 1

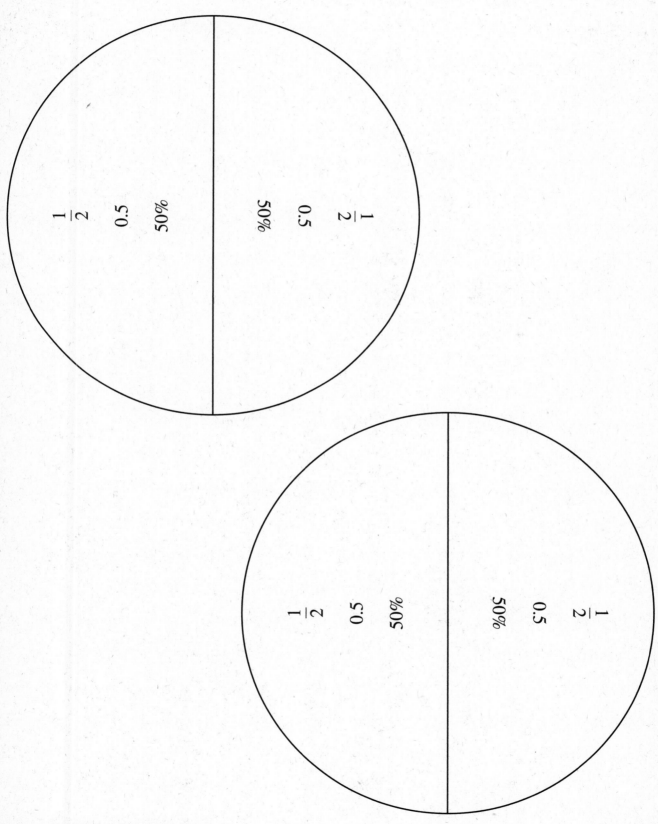

2 | Thirds
For use with Investigation 1

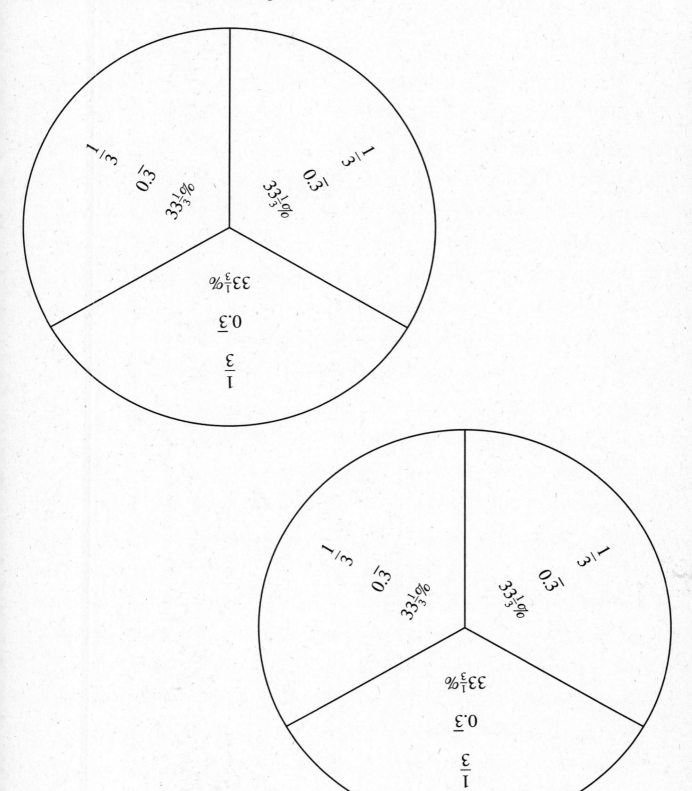

3 Fourths
For use with Investigation 1

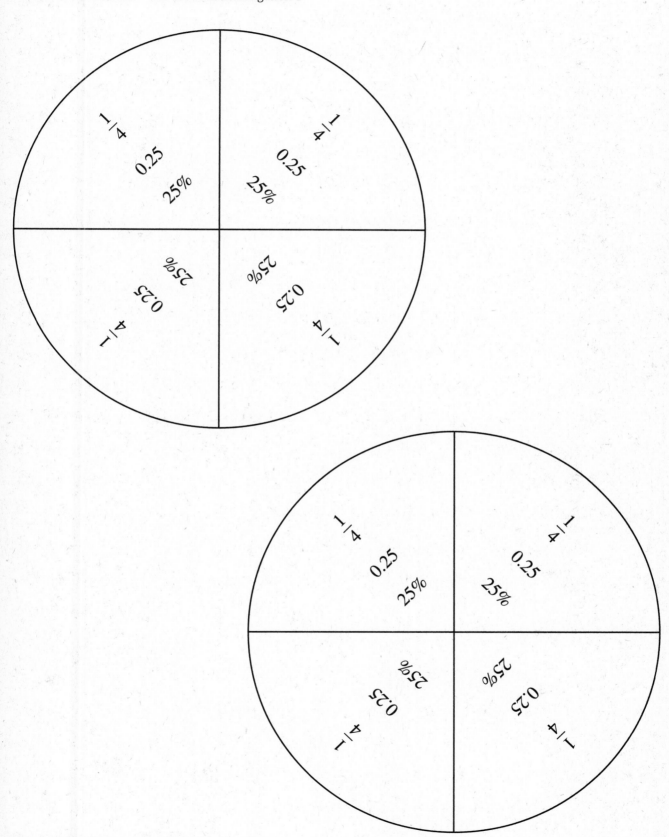

4 | Sixths

For use with Investigation 1

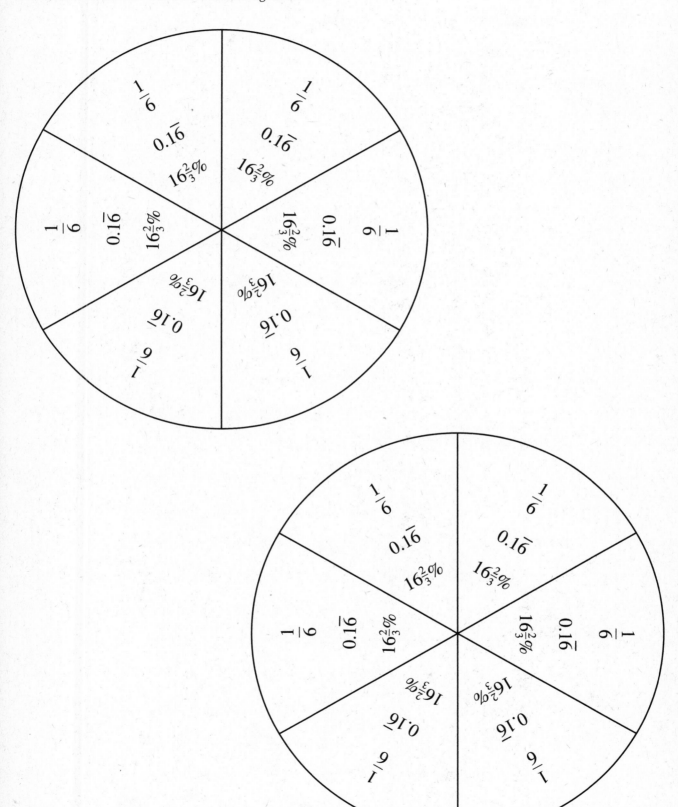

5 Eighths

For use with Investigation 1

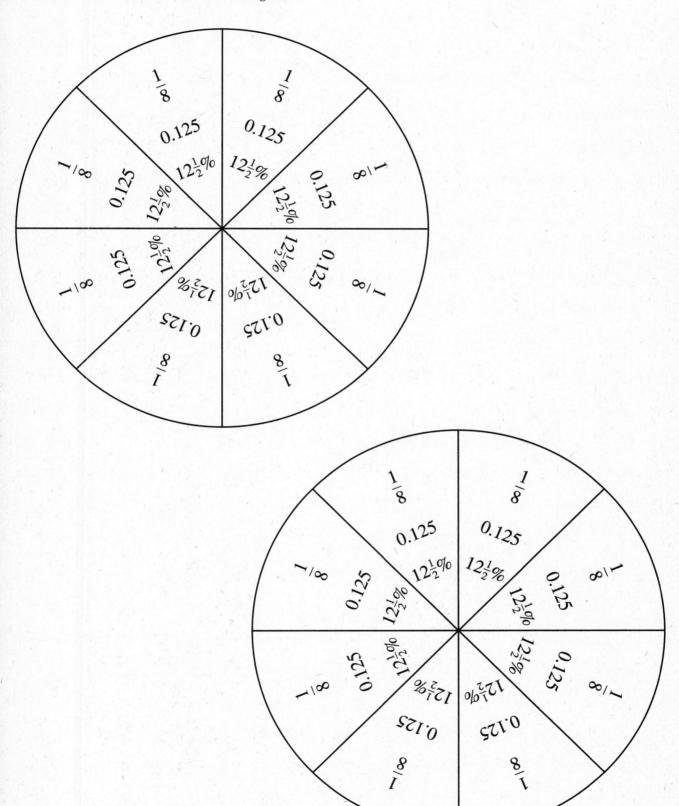

6   Twelfths

For use with Investigation 1

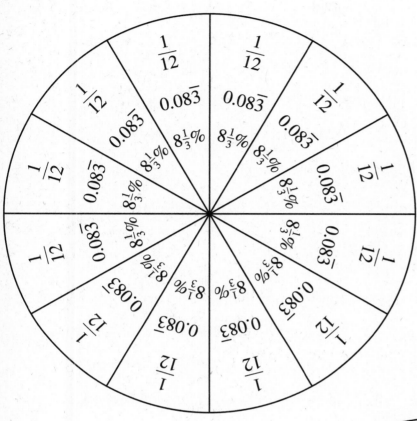

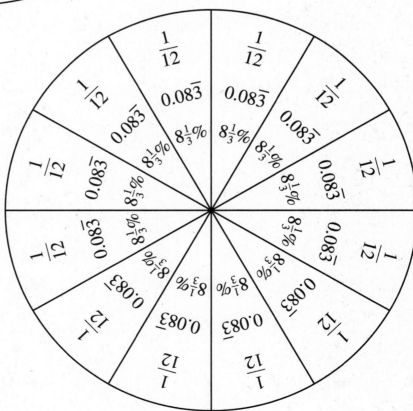

C

30 Improper Fractions and Mixed Numbers

For use with Lesson 11

Name _____

Time _____

Write each improper fraction as a mixed number or a whole number.

$\frac{5}{2}$ =	$\frac{6}{3}$ =	$\frac{7}{4}$ =	$\frac{12}{5}$ =	$\frac{8}{2}$ =
$\frac{10}{3}$ =	$\frac{15}{2}$ =	$\frac{21}{4}$ =	$\frac{15}{5}$ =	$\frac{11}{8}$ =
$2\frac{3}{2}$ =	$4\frac{5}{4}$ =	$3\frac{6}{2}$ =	$3\frac{7}{4}$ =	$6\frac{5}{2}$ =

Write each mixed number as an improper fraction.

$1\frac{1}{2}$ =	$2\frac{2}{3}$ =	$3\frac{3}{4}$ =	$2\frac{1}{2}$ =	$4\frac{1}{5}$ =
$6\frac{2}{3}$ =	$2\frac{3}{4}$ =	$3\frac{1}{3}$ =	$4\frac{1}{2}$ =	$2\frac{4}{5}$ =
$1\frac{5}{6}$ =	$5\frac{3}{4}$ =	$1\frac{7}{8}$ =	$3\frac{1}{6}$ =	$2\frac{3}{10}$ =

C

30 Improper Fractions and Mixed Numbers
For use with Lesson 12

Name _____

Time _____

Write each improper fraction as a mixed number or a whole number.

$\frac{5}{2} =$	$\frac{6}{3} =$	$\frac{7}{4} =$	$\frac{12}{5} =$	$\frac{8}{2} =$
$\frac{10}{3} =$	$\frac{15}{2} =$	$\frac{21}{4} =$	$\frac{15}{5} =$	$\frac{11}{8} =$
$2\frac{3}{2} =$	$4\frac{5}{4} =$	$3\frac{6}{2} =$	$3\frac{7}{4} =$	$6\frac{5}{2} =$

Write each mixed number as an improper fraction.

$1\frac{1}{2} =$	$2\frac{2}{3} =$	$3\frac{3}{4} =$	$2\frac{1}{2} =$	$4\frac{1}{5} =$
$6\frac{2}{3} =$	$2\frac{3}{4} =$	$3\frac{1}{3} =$	$4\frac{1}{2} =$	$2\frac{4}{5} =$
$1\frac{5}{6} =$	$5\frac{3}{4} =$	$1\frac{7}{8} =$	$3\frac{1}{6} =$	$2\frac{3}{10} =$

C | 30 Improper Fractions and Mixed Numbers

For use with Lesson 13

Name _____

Time _____

Write each improper fraction as a mixed number or a whole number.

$\frac{5}{2} =$	$\frac{6}{3} =$	$\frac{7}{4} =$	$\frac{12}{5} =$	$\frac{8}{2} =$
$\frac{10}{3} =$	$\frac{15}{2} =$	$\frac{21}{4} =$	$\frac{15}{5} =$	$\frac{11}{8} =$
$2\frac{3}{2} =$	$4\frac{5}{4} =$	$3\frac{6}{2} =$	$3\frac{7}{4} =$	$6\frac{5}{2} =$

Write each mixed number as an improper fraction.

$1\frac{1}{2} =$	$2\frac{2}{3} =$	$3\frac{3}{4} =$	$2\frac{1}{2} =$	$4\frac{1}{5} =$
$6\frac{2}{3} =$	$2\frac{3}{4} =$	$3\frac{1}{3} =$	$4\frac{1}{2} =$	$2\frac{4}{5} =$
$1\frac{5}{6} =$	$5\frac{3}{4} =$	$1\frac{7}{8} =$	$3\frac{1}{6} =$	$2\frac{3}{10} =$

B 30 Equations
For use with Lesson 14

Name _____

Time _____

Find the value of each variable.

$a + 12 = 20$ $a =$	$b - 8 = 10$ $b =$	$5c = 40$ $c =$
$\dfrac{d}{4} = 12$ $d =$	$11 + e = 24$ $e =$	$25 - f = 10$ $f =$
$10g = 60$ $g =$	$\dfrac{24}{h} = 6$ $h =$	$17 = j + 8$ $j =$
$20 = k - 5$ $k =$	$30 = 6m$ $m =$	$9 = \dfrac{n}{3}$ $n =$
$18 = 6 + p$ $p =$	$5 = 15 - q$ $q =$	$36 = 4r$ $r =$
$2 = \dfrac{16}{s}$ $s =$	$5 + 7 + t = 20$ $t =$	$u - 15 = 30$ $u =$
$8v = 48$ $v =$	$\dfrac{w}{3} = 6$ $w =$	$21 - x = 12$ $x =$
$y + 8 = 12$ $y =$	$36 = 3z$ $z =$	$\dfrac{48}{a} = 4$ $a =$
$b - 12 = 15$ $b =$	$75 = 3c$ $c =$	$\dfrac{d}{12} = 6$ $d =$
$36 = f + 24$ $f =$	$g - 24 = 24$ $g =$	$12h = 12$ $h =$

Saxon Math 8/7—Homeschool

30 Improper Fractions and Mixed Numbers

For use with Lesson 15

Name _____

Time _____

Write each improper fraction as a mixed number or a whole number.

$\frac{5}{2}$ =	$\frac{6}{3}$ =	$\frac{7}{4}$ =	$\frac{12}{5}$ =	$\frac{8}{2}$ =
$\frac{10}{3}$ =	$\frac{15}{2}$ =	$\frac{21}{4}$ =	$\frac{15}{5}$ =	$\frac{11}{8}$ =
$2\frac{3}{2}$ =	$4\frac{5}{4}$ =	$3\frac{6}{2}$ =	$3\frac{7}{4}$ =	$6\frac{5}{2}$ =

Write each mixed number as an improper fraction.

$1\frac{1}{2}$ =	$2\frac{2}{3}$ =	$3\frac{3}{4}$ =	$2\frac{1}{2}$ =	$4\frac{1}{5}$ =
$6\frac{2}{3}$ =	$2\frac{3}{4}$ =	$3\frac{1}{3}$ =	$4\frac{1}{2}$ =	$2\frac{4}{5}$ =
$1\frac{5}{6}$ =	$5\frac{3}{4}$ =	$1\frac{7}{8}$ =	$3\frac{1}{6}$ =	$2\frac{3}{10}$ =

B

30 Equations
For use with Test 2

Name _____

Time _____

Find the value of each variable.

$a + 12 = 20$ $a =$	$b - 8 = 10$ $b =$	$5c = 40$ $c =$
$\dfrac{d}{4} = 12$ $d =$	$11 + e = 24$ $e =$	$25 - f = 10$ $f =$
$10g = 60$ $g =$	$\dfrac{24}{h} = 6$ $h =$	$17 = j + 8$ $j =$
$20 = k - 5$ $k =$	$30 = 6m$ $m =$	$9 = \dfrac{n}{3}$ $n =$
$18 = 6 + p$ $p =$	$5 = 15 - q$ $q =$	$36 = 4r$ $r =$
$2 = \dfrac{16}{s}$ $s =$	$5 + 7 + t = 20$ $t =$	$u - 15 = 30$ $u =$
$8v = 48$ $v =$	$\dfrac{w}{3} = 6$ $w =$	$21 - x = 12$ $x =$
$y + 8 = 12$ $y =$	$36 = 3z$ $z =$	$\dfrac{48}{a} = 4$ $a =$
$b - 12 = 15$ $b =$	$75 = 3c$ $c =$	$\dfrac{d}{12} = 6$ $d =$
$36 = f + 24$ $f =$	$g - 24 = 24$ $g =$	$12h = 12$ $h =$

D | 40 Fractions to Reduce

For use with Lesson 16

Name _____

Time _____

Reduce each fraction to lowest terms.

$\frac{60}{100}=$	$\frac{2}{12}=$	$\frac{4}{16}=$	$\frac{2}{6}=$	$\frac{5}{10}=$
$\frac{50}{100}=$	$\frac{2}{16}=$	$\frac{8}{12}=$	$\frac{5}{100}=$	$\frac{3}{9}=$
$\frac{8}{16}=$	$\frac{2}{100}=$	$\frac{20}{100}=$	$\frac{6}{8}=$	$\frac{10}{100}=$
$\frac{2}{4}=$	$\frac{4}{10}=$	$\frac{90}{100}=$	$\frac{3}{12}=$	$\frac{6}{16}=$
$\frac{80}{100}=$	$\frac{9}{12}=$	$\frac{3}{6}=$	$\frac{12}{16}=$	$\frac{4}{8}=$
$\frac{6}{9}=$	$\frac{25}{100}=$	$\frac{4}{12}=$	$\frac{6}{10}=$	$\frac{40}{100}=$
$\frac{4}{100}=$	$\frac{2}{10}=$	$\frac{10}{16}=$	$\frac{10}{12}=$	$\frac{4}{6}=$
$\frac{14}{16}=$	$\frac{2}{8}=$	$\frac{6}{12}=$	$\frac{8}{10}=$	$\frac{75}{100}=$

D 40 Fractions to Reduce
For use with Lesson 17

Name _____

Time _____

Reduce each fraction to lowest terms.

$\frac{60}{100} =$	$\frac{2}{12} =$	$\frac{4}{16} =$	$\frac{2}{6} =$	$\frac{5}{10} =$
$\frac{50}{100} =$	$\frac{2}{16} =$	$\frac{8}{12} =$	$\frac{5}{100} =$	$\frac{3}{9} =$
$\frac{8}{16} =$	$\frac{2}{100} =$	$\frac{20}{100} =$	$\frac{6}{8} =$	$\frac{10}{100} =$
$\frac{2}{4} =$	$\frac{4}{10} =$	$\frac{90}{100} =$	$\frac{3}{12} =$	$\frac{6}{16} =$
$\frac{80}{100} =$	$\frac{9}{12} =$	$\frac{3}{6} =$	$\frac{12}{16} =$	$\frac{4}{8} =$
$\frac{6}{9} =$	$\frac{25}{100} =$	$\frac{4}{12} =$	$\frac{6}{10} =$	$\frac{40}{100} =$
$\frac{4}{100} =$	$\frac{2}{10} =$	$\frac{10}{16} =$	$\frac{10}{12} =$	$\frac{4}{6} =$
$\frac{14}{16} =$	$\frac{2}{8} =$	$\frac{6}{12} =$	$\frac{8}{10} =$	$\frac{75}{100} =$

Saxon Math 8/7—Homeschool

D   **40 Fractions to Reduce**
For use with Lesson 18

Name _____

Time _____

Reduce each fraction to lowest terms.

$\frac{60}{100} =$	$\frac{2}{12} =$	$\frac{4}{16} =$	$\frac{2}{6} =$	$\frac{5}{10} =$
$\frac{50}{100} =$	$\frac{2}{16} =$	$\frac{8}{12} =$	$\frac{5}{100} =$	$\frac{3}{9} =$
$\frac{8}{16} =$	$\frac{2}{100} =$	$\frac{20}{100} =$	$\frac{6}{8} =$	$\frac{10}{100} =$
$\frac{2}{4} =$	$\frac{4}{10} =$	$\frac{90}{100} =$	$\frac{3}{12} =$	$\frac{6}{16} =$
$\frac{80}{100} =$	$\frac{9}{12} =$	$\frac{3}{6} =$	$\frac{12}{16} =$	$\frac{4}{8} =$
$\frac{6}{9} =$	$\frac{25}{100} =$	$\frac{4}{12} =$	$\frac{6}{10} =$	$\frac{40}{100} =$
$\frac{4}{100} =$	$\frac{2}{10} =$	$\frac{10}{16} =$	$\frac{10}{12} =$	$\frac{4}{6} =$
$\frac{14}{16} =$	$\frac{2}{8} =$	$\frac{6}{12} =$	$\frac{8}{10} =$	$\frac{75}{100} =$

C 30 Improper Fractions and Mixed Numbers

For use with Lesson 19

Name _____

Time _____

Write each improper fraction as a mixed number or a whole number.

$\dfrac{5}{2} =$	$\dfrac{6}{3} =$	$\dfrac{7}{4} =$	$\dfrac{12}{5} =$	$\dfrac{8}{2} =$
$\dfrac{10}{3} =$	$\dfrac{15}{2} =$	$\dfrac{21}{4} =$	$\dfrac{15}{5} =$	$\dfrac{11}{8} =$
$2\dfrac{3}{2} =$	$4\dfrac{5}{4} =$	$3\dfrac{6}{2} =$	$3\dfrac{7}{4} =$	$6\dfrac{5}{2} =$

Write each mixed number as an improper fraction.

$1\dfrac{1}{2} =$	$2\dfrac{2}{3} =$	$3\dfrac{3}{4} =$	$2\dfrac{1}{2} =$	$4\dfrac{1}{5} =$
$6\dfrac{2}{3} =$	$2\dfrac{3}{4} =$	$3\dfrac{1}{3} =$	$4\dfrac{1}{2} =$	$2\dfrac{4}{5} =$
$1\dfrac{5}{6} =$	$5\dfrac{3}{4} =$	$1\dfrac{7}{8} =$	$3\dfrac{1}{6} =$	$2\dfrac{3}{10} =$

D

40 Fractions to Reduce
For use with Lesson 20

Name _____

Time _____

Reduce each fraction to lowest terms.

$\frac{60}{100} =$	$\frac{2}{12} =$	$\frac{4}{16} =$	$\frac{2}{6} =$	$\frac{5}{10} =$
$\frac{50}{100} =$	$\frac{2}{16} =$	$\frac{8}{12} =$	$\frac{5}{100} =$	$\frac{3}{9} =$
$\frac{8}{16} =$	$\frac{2}{100} =$	$\frac{20}{100} =$	$\frac{6}{8} =$	$\frac{10}{100} =$
$\frac{2}{4} =$	$\frac{4}{10} =$	$\frac{90}{100} =$	$\frac{3}{12} =$	$\frac{6}{16} =$
$\frac{80}{100} =$	$\frac{9}{12} =$	$\frac{3}{6} =$	$\frac{12}{16} =$	$\frac{4}{8} =$
$\frac{6}{9} =$	$\frac{25}{100} =$	$\frac{4}{12} =$	$\frac{6}{10} =$	$\frac{40}{100} =$
$\frac{4}{100} =$	$\frac{2}{10} =$	$\frac{10}{16} =$	$\frac{10}{12} =$	$\frac{4}{6} =$
$\frac{14}{16} =$	$\frac{2}{8} =$	$\frac{6}{12} =$	$\frac{8}{10} =$	$\frac{75}{100} =$

C

30 Improper Fractions and Mixed Numbers
For use with Test 3

Name _____

Time _____

Write each improper fraction as a mixed number or a whole number.

$\frac{5}{2} =$	$\frac{6}{3} =$	$\frac{7}{4} =$	$\frac{12}{5} =$	$\frac{8}{2} =$
$\frac{10}{3} =$	$\frac{15}{2} =$	$\frac{21}{4} =$	$\frac{15}{5} =$	$\frac{11}{8} =$
$2\frac{3}{2} =$	$4\frac{5}{4} =$	$3\frac{6}{2} =$	$3\frac{7}{4} =$	$6\frac{5}{2} =$

Write each mixed number as an improper fraction.

$1\frac{1}{2} =$	$2\frac{2}{3} =$	$3\frac{3}{4} =$	$2\frac{1}{2} =$	$4\frac{1}{5} =$
$6\frac{2}{3} =$	$2\frac{3}{4} =$	$3\frac{1}{3} =$	$4\frac{1}{2} =$	$2\frac{4}{5} =$
$1\frac{5}{6} =$	$5\frac{3}{4} =$	$1\frac{7}{8} =$	$3\frac{1}{6} =$	$2\frac{3}{10} =$

E	**Circles** *For use with Lesson 21*

Name _____

Time _____

Write the word that completes each sentence.

1. The distance around a circle is its _____.

2. Every point on a circle is the same distance from the _____.

3. The distance across a circle through its center is its _____.

4. The distance from a circle to its center is its _____.

5. Two or more circles with the same center are _____ circles.

6. A segment between two points on a circle is a _____.

7. Part of a circumference is an _____.

8. A portion of a circle and its interior, bound by an arc and two radii, is a _____.

9. Half of a circle is a _____.

10. An angle whose vertex is the center of a circle is a _____ angle.

11. An angle whose vertex is on the circumference of a circle and whose sides include chords of the circle is an _____ angle.

12. A polygon whose vertices are on a circle and whose edges are within the circle is an _____ polygon.

Illustrate answers 1–12 below.

1.	2.	3.	4.
5.	**6.**	**7.**	**8.**
9.	**10.**	**11.**	**12.**

| E |

Circles
For use with Lesson 22

Name _____

Time _____

Write the word that completes each sentence.

1. The distance around a circle is its _____.

2. Every point on a circle is the same distance from the _____.

3. The distance across a circle through its center is its _____.

4. The distance from a circle to its center is its _____.

5. Two or more circles with the same center are _____ circles.

6. A segment between two points on a circle is a _____.

7. Part of a circumference is an _____.

8. A portion of a circle and its interior, bound by an arc and two radii, is a _____.

9. Half of a circle is a _____.

10. An angle whose vertex is the center of a circle is a _____ angle.

11. An angle whose vertex is on the circumference of a circle and whose sides include chords of the circle is an _____ angle.

12. A polygon whose vertices are on a circle and whose edges are within the circle is an _____ polygon.

Illustrate answers 1–12 below.

1.	2.	3.	4.
5.	**6.**	**7.**	**8.**
9.	**10.**	**11.**	**12.**

E	**Circles** *For use with Lesson 23*

Name _____

Time _____

Write the word that completes each sentence.

1. The distance around a circle is its _____.

2. Every point on a circle is the same distance from the _____.

3. The distance across a circle through its center is its _____.

4. The distance from a circle to its center is its _____.

5. Two or more circles with the same center are _____ circles.

6. A segment between two points on a circle is a _____.

7. Part of a circumference is an _____.

8. A portion of a circle and its interior, bound by an arc and two radii, is a _____.

9. Half of a circle is a _____.

10. An angle whose vertex is the center of a circle is a _____ angle.

11. An angle whose vertex is on the circumference of a circle and whose sides include chords of the circle is an _____ angle.

12. A polygon whose vertices are on a circle and whose edges are within the circle is an _____ polygon.

Illustrate answers 1–12 below.

1.	2.	3.	4.
5.	**6.**	**7.**	**8.**
9.	**10.**	**11.**	**12.**

D **40 Fractions to Reduce**
For use with Lesson 24

Name _____

Time _____

Reduce each fraction to lowest terms.

$\dfrac{60}{100} =$	$\dfrac{2}{12} =$	$\dfrac{4}{16} =$	$\dfrac{2}{6} =$	$\dfrac{5}{10} =$
$\dfrac{50}{100} =$	$\dfrac{2}{16} =$	$\dfrac{8}{12} =$	$\dfrac{5}{100} =$	$\dfrac{3}{9} =$
$\dfrac{8}{16} =$	$\dfrac{2}{100} =$	$\dfrac{20}{100} =$	$\dfrac{6}{8} =$	$\dfrac{10}{100} =$
$\dfrac{2}{4} =$	$\dfrac{4}{10} =$	$\dfrac{90}{100} =$	$\dfrac{3}{12} =$	$\dfrac{6}{16} =$
$\dfrac{80}{100} =$	$\dfrac{9}{12} =$	$\dfrac{3}{6} =$	$\dfrac{12}{16} =$	$\dfrac{4}{8} =$
$\dfrac{6}{9} =$	$\dfrac{25}{100} =$	$\dfrac{4}{12} =$	$\dfrac{6}{10} =$	$\dfrac{40}{100} =$
$\dfrac{4}{100} =$	$\dfrac{2}{10} =$	$\dfrac{10}{16} =$	$\dfrac{10}{12} =$	$\dfrac{4}{6} =$
$\dfrac{14}{16} =$	$\dfrac{2}{8} =$	$\dfrac{6}{12} =$	$\dfrac{8}{10} =$	$\dfrac{75}{100} =$

Saxon Math 8/7—Homeschool

F **Lines, Angles, Polygons**
For use with Lesson 25

Name _____

Time _____

Name each figure illustrated.

1.	**2.**	**3.**
_____	_____	_____
4.	**5.**	**6.**
_____	_____	_____
7.	**8.**	**9.**
_____	_____	_____
10.	**11.**	**12.**
_____	_____	_____
13.	**14.**	**15.**
_____	_____	_____

16. A polygon whose sides are equal in length and whose angles are equal in measure is a
_____ polygon.

D

40 Fractions to Reduce

For use with Test 4

Name _____

Time _____

Reduce each fraction to lowest terms.

$\frac{60}{100} =$	$\frac{2}{12} =$	$\frac{4}{16} =$	$\frac{2}{6} =$	$\frac{5}{10} =$
$\frac{50}{100} =$	$\frac{2}{16} =$	$\frac{8}{12} =$	$\frac{5}{100} =$	$\frac{3}{9} =$
$\frac{8}{16} =$	$\frac{2}{100} =$	$\frac{20}{100} =$	$\frac{6}{8} =$	$\frac{10}{100} =$
$\frac{2}{4} =$	$\frac{4}{10} =$	$\frac{90}{100} =$	$\frac{3}{12} =$	$\frac{6}{16} =$
$\frac{80}{100} =$	$\frac{9}{12} =$	$\frac{3}{6} =$	$\frac{12}{16} =$	$\frac{4}{8} =$
$\frac{6}{9} =$	$\frac{25}{100} =$	$\frac{4}{12} =$	$\frac{6}{10} =$	$\frac{40}{100} =$
$\frac{4}{100} =$	$\frac{2}{10} =$	$\frac{10}{16} =$	$\frac{10}{12} =$	$\frac{4}{6} =$
$\frac{14}{16} =$	$\frac{2}{8} =$	$\frac{6}{12} =$	$\frac{8}{10} =$	$\frac{75}{100} =$

F | Lines, Angles, Polygons
For use with Lesson 26

Name _____

Time _____

Name each figure illustrated.

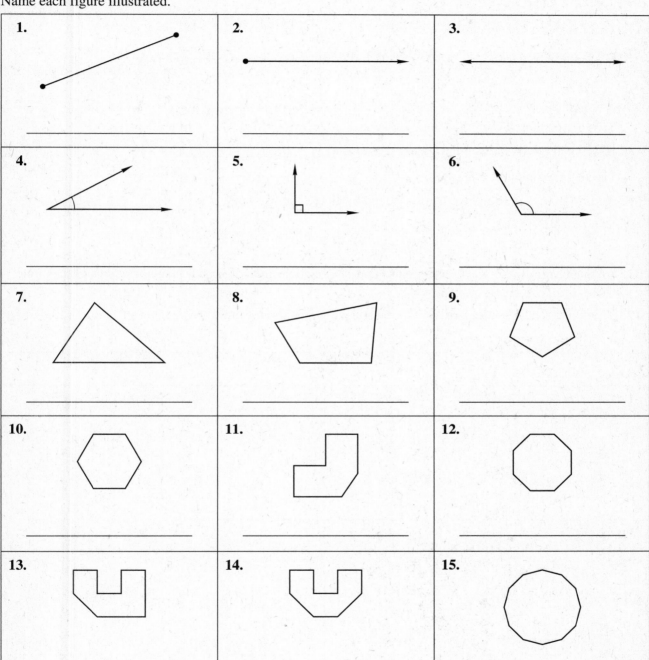

1. _____

2. _____

3. _____

4. _____

5. _____

6. _____

7. _____

8. _____

9. _____

10. _____

11. _____

12. _____

13. _____

14. _____

15. _____

16. A polygon whose sides are equal in length and whose angles are equal in measure is a
_____ polygon.

E | Circles
For use with Lesson 27

Name _____

Time _____

Write the word that completes each sentence.

1. The distance around a circle is its _____.

2. Every point on a circle is the same distance from the _____.

3. The distance across a circle through its center is its _____.

4. The distance from a circle to its center is its _____.

5. Two or more circles with the same center are _____ circles.

6. A segment between two points on a circle is a _____.

7. Part of a circumference is an _____.

8. A portion of a circle and its interior, bound by an arc and two radii, is a _____.

9. Half of a circle is a _____.

10. An angle whose vertex is the center of a circle is a _____ angle.

11. An angle whose vertex is on the circumference of a circle and whose sides include chords of the circle is an _____ angle.

12. A polygon whose vertices are on a circle and whose edges are within the circle is an _____ polygon.

Illustrate answers 1–12 below.

1.	2.	3.	4.
5.	6.	7.	8.
9.	10.	11.	12.

Saxon Math 8/7—Homeschool

| F |

Lines, Angles, Polygons
For use with Lesson 28

Name _____

Time _____

Name each figure illustrated.

1.	2.	3.
_____	_____	_____
4.	5.	6.
_____	_____	_____
7.	8.	9.
_____	_____	_____
10.	11.	12.
_____	_____	_____
13.	14.	15.
_____	_____	_____

16. A polygon whose sides are equal in length and whose angles are equal in measure is a
_____ polygon.

E	**Circles**	Name _____
	For use with Lesson 29	Time _____

Write the word that completes each sentence.

1. The distance around a circle is its _____.

2. Every point on a circle is the same distance from the _____.

3. The distance across a circle through its center is its _____.

4. The distance from a circle to its center is its _____.

5. Two or more circles with the same center are _____ circles.

6. A segment between two points on a circle is a _____.

7. Part of a circumference is an _____.

8. A portion of a circle and its interior, bound by an arc and two radii, is a _____.

9. Half of a circle is a _____.

10. An angle whose vertex is the center of a circle is a _____ angle.

11. An angle whose vertex is on the circumference of a circle and whose sides include chords of the circle is an _____ angle.

12. A polygon whose vertices are on a circle and whose edges are within the circle is an _____ polygon.

Illustrate answers 1–12 below.

1.	2.	3.	4.
5.	**6.**	**7.**	**8.**
9.	**10.**	**11.**	**12.**

F

Lines, Angles, Polygons
For use with Lesson 30

Name _____

Time _____

Name each figure illustrated.

1.	**2.**	**3.**
_____	_____	_____
4.	**5.**	**6.**
_____	_____	_____
7.	**8.**	**9.**
_____	_____	_____
10.	**11.**	**12.**
_____	_____	_____
13.	**14.**	**15.**
_____	_____	_____

16. A polygon whose sides are equal in length and whose angles are equal in measure is a
_____ polygon.

E	**Circles**

For use with Test 5

Name _____

Time _____

Write the word that completes each sentence.

1. The distance around a circle is its _____.

2. Every point on a circle is the same distance from the _____.

3. The distance across a circle through its center is its _____.

4. The distance from a circle to its center is its _____.

5. Two or more circles with the same center are _____ circles.

6. A segment between two points on a circle is a _____.

7. Part of a circumference is an _____.

8. A portion of a circle and its interior, bound by an arc and two radii, is a _____.

9. Half of a circle is a _____.

10. An angle whose vertex is the center of a circle is a _____ angle.

11. An angle whose vertex is on the circumference of a circle and whose sides include chords of the circle is an _____ angle.

12. A polygon whose vertices are on a circle and whose edges are within the circle is an _____ polygon.

Illustrate answers 1–12 below.

1.	2.	3.	4.
5.	**6.**	**7.**	**8.**
9.	**10.**	**11.**	**12.**

G | **+ − × ÷ Fractions**
For use with Lesson 31

Name _____

Time _____

Simplify these expressions. Reduce the answers.

$\frac{2}{3} + \frac{2}{3} =$	$\frac{2}{3} - \frac{2}{3} =$	$\frac{2}{3} \times \frac{2}{3} =$	$\frac{2}{3} \div \frac{2}{3} =$
$\frac{3}{4} + \frac{1}{4} =$	$\frac{3}{4} - \frac{1}{4} =$	$\frac{3}{4} \times \frac{1}{4} =$	$\frac{3}{4} \div \frac{1}{4} =$
$\frac{2}{3} + \frac{1}{2} =$	$\frac{2}{3} - \frac{1}{2} =$	$\frac{2}{3} \times \frac{1}{2} =$	$\frac{2}{3} \div \frac{1}{2} =$
$\frac{3}{4} + \frac{2}{3} =$	$\frac{3}{4} - \frac{2}{3} =$	$\frac{3}{4} \times \frac{2}{3} =$	$\frac{3}{4} \div \frac{2}{3} =$
$\frac{2}{5} + \frac{1}{4} =$	$\frac{2}{5} - \frac{1}{4} =$	$\frac{2}{5} \times \frac{1}{4} =$	$\frac{2}{5} \div \frac{1}{4} =$
$\frac{1}{2} + \frac{5}{8} =$	$\frac{5}{8} - \frac{1}{2} =$	$\frac{1}{2} \times \frac{5}{8} =$	$\frac{1}{2} \div \frac{5}{8} =$

G

+ − × ÷ Fractions
For use with Lesson 32

Name _____

Time _____

Simplify these expressions. Reduce the answers.

$\frac{2}{3} + \frac{2}{3} =$	$\frac{2}{3} - \frac{2}{3} =$	$\frac{2}{3} \times \frac{2}{3} =$	$\frac{2}{3} \div \frac{2}{3} =$
$\frac{3}{4} + \frac{1}{4} =$	$\frac{3}{4} - \frac{1}{4} =$	$\frac{3}{4} \times \frac{1}{4} =$	$\frac{3}{4} \div \frac{1}{4} =$
$\frac{2}{3} + \frac{1}{2} =$	$\frac{2}{3} - \frac{1}{2} =$	$\frac{2}{3} \times \frac{1}{2} =$	$\frac{2}{3} \div \frac{1}{2} =$
$\frac{3}{4} + \frac{2}{3} =$	$\frac{3}{4} - \frac{2}{3} =$	$\frac{3}{4} \times \frac{2}{3} =$	$\frac{3}{4} \div \frac{2}{3} =$
$\frac{2}{5} + \frac{1}{4} =$	$\frac{2}{5} - \frac{1}{4} =$	$\frac{2}{5} \times \frac{1}{4} =$	$\frac{2}{5} \div \frac{1}{4} =$
$\frac{1}{2} + \frac{5}{8} =$	$\frac{5}{8} - \frac{1}{2} =$	$\frac{1}{2} \times \frac{5}{8} =$	$\frac{1}{2} \div \frac{5}{8} =$

F | Lines, Angles, Polygons

For use with Lesson 33

Name _____

Time _____

Name each figure illustrated.

1.	2.	3.
_____	_____	_____
4.	5.	6.
_____	_____	_____
7.	8.	9.
_____	_____	_____
10.	11.	12.
_____	_____	_____
13.	14.	15.
_____	_____	_____

16. A polygon whose sides are equal in length and whose angles are equal in measure is a
_____ polygon.

G

$+ - \times \div$ Fractions
For use with Lesson 34

Name _____

Time _____

Simplify these expressions. Reduce the answers.

$\dfrac{2}{3} + \dfrac{2}{3} =$	$\dfrac{2}{3} - \dfrac{2}{3} =$	$\dfrac{2}{3} \times \dfrac{2}{3} =$	$\dfrac{2}{3} \div \dfrac{2}{3} =$
$\dfrac{3}{4} + \dfrac{1}{4} =$	$\dfrac{3}{4} - \dfrac{1}{4} =$	$\dfrac{3}{4} \times \dfrac{1}{4} =$	$\dfrac{3}{4} \div \dfrac{1}{4} =$
$\dfrac{2}{3} + \dfrac{1}{2} =$	$\dfrac{2}{3} - \dfrac{1}{2} =$	$\dfrac{2}{3} \times \dfrac{1}{2} =$	$\dfrac{2}{3} \div \dfrac{1}{2} =$
$\dfrac{3}{4} + \dfrac{2}{3} =$	$\dfrac{3}{4} - \dfrac{2}{3} =$	$\dfrac{3}{4} \times \dfrac{2}{3} =$	$\dfrac{3}{4} \div \dfrac{2}{3} =$
$\dfrac{2}{5} + \dfrac{1}{4} =$	$\dfrac{2}{5} - \dfrac{1}{4} =$	$\dfrac{2}{5} \times \dfrac{1}{4} =$	$\dfrac{2}{5} \div \dfrac{1}{4} =$
$\dfrac{1}{2} + \dfrac{5}{8} =$	$\dfrac{5}{8} - \dfrac{1}{2} =$	$\dfrac{1}{2} \times \dfrac{5}{8} =$	$\dfrac{1}{2} \div \dfrac{5}{8} =$

H Measurement Facts
For use with Lesson 35

Name _____

Time _____

Write the number that makes each statement true.

Customary Units

Linear Measure

1. 1 foot = _____ inches

2. 1 yard = _____ inches

3. 1 yard = _____ feet

4. 1 mile = _____ feet

5. 1 mile = _____ yards

Area

6. 1 foot^2 = _____ inches^2

7. 1 yard^2 = _____ feet^2

Volume

8. 1 yard^3 = _____ feet^3

Weight

9. 1 pound = _____ ounces

10. 1 ton = _____ pounds

Liquid Measure

11. 1 pint = _____ ounces

12. 1 pint = _____ cups

13. 1 quart = _____ pints

14. 1 gallon = _____ quarts

Temperature

15. Water freezes at _____ °F.

16. Water boils at _____ °F.

17. Normal body temperature is _____ °F.

Customary to Metric

18. 1 inch = _____ centimeters

Metric Units

Linear Measure

19. 1 centimeter = _____ millimeters

20. 1 meter = _____ centimeters

21. 1 meter = _____ millimeters

22. 1 kilometer = _____ meters

Area

23. 1 meter^2 = _____ centimeters^2

24. 1 kilometer^2 = _____ meters^2

Volume

25. 1 meter^3 = _____ centimeters^3

Mass

26. 1 gram = _____ milligrams

27. 1 kilogram = _____ grams

28. 1 metric ton = _____ kilograms

Capacity

29. 1 liter = _____ milliliters

30. One milliliter of water has a volume of _____ and a mass of _____. One thousand cm^3 of water fills a _____-liter container and has a mass of _____ kilogram.

Temperature

31. Water freezes at _____ °C.

32. Water boils at _____ °C.

33. Normal body temperature is _____ °C.

F | Lines, Angles, Polygons
For use with Test 6

Name _____

Time _____

Name each figure illustrated.

1.	**2.**	**3.**
_____	_____	_____
4.	**5.**	**6.**
_____	_____	_____
7.	**8.**	**9.**
_____	_____	_____
10.	**11.**	**12.**
_____	_____	_____
13.	**14.**	**15.**
_____	_____	_____

16. A polygon whose sides are equal in length and whose angles are equal in measure is a _____ polygon.

Saxon Math 8/7—Homeschool

G	$+ - \times \div$ **Fractions**

For use with Lesson 36

Name _____

Time _____

Simplify these expressions. Reduce the answers.

$\frac{2}{3} + \frac{2}{3} =$	$\frac{2}{3} - \frac{2}{3} =$	$\frac{2}{3} \times \frac{2}{3} =$	$\frac{2}{3} \div \frac{2}{3} =$
$\frac{3}{4} + \frac{1}{4} =$	$\frac{3}{4} - \frac{1}{4} =$	$\frac{3}{4} \times \frac{1}{4} =$	$\frac{3}{4} \div \frac{1}{4} =$
$\frac{2}{3} + \frac{1}{2} =$	$\frac{2}{3} - \frac{1}{2} =$	$\frac{2}{3} \times \frac{1}{2} =$	$\frac{2}{3} \div \frac{1}{2} =$
$\frac{3}{4} + \frac{2}{3} =$	$\frac{3}{4} - \frac{2}{3} =$	$\frac{3}{4} \times \frac{2}{3} =$	$\frac{3}{4} \div \frac{2}{3} =$
$\frac{2}{5} + \frac{1}{4} =$	$\frac{2}{5} - \frac{1}{4} =$	$\frac{2}{5} \times \frac{1}{4} =$	$\frac{2}{5} \div \frac{1}{4} =$
$\frac{1}{2} + \frac{5}{8} =$	$\frac{5}{8} - \frac{1}{2} =$	$\frac{1}{2} \times \frac{5}{8} =$	$\frac{1}{2} \div \frac{5}{8} =$

H Measurement Facts
For use with Lesson 37

Name _____

Time _____

Write the number that makes each statement true.

Customary Units

Linear Measure

1. 1 foot = _____ inches

2. 1 yard = _____ inches

3. 1 yard = _____ feet

4. 1 mile = _____ feet

5. 1 mile = _____ yards

Area

6. 1 foot^2 = _____ inches^2

7. 1 yard^2 = _____ feet^2

Volume

8. 1 yard^3 = _____ feet^3

Weight

9. 1 pound = _____ ounces

10. 1 ton = _____ pounds

Liquid Measure

11. 1 pint = _____ ounces

12. 1 pint = _____ cups

13. 1 quart = _____ pints

14. 1 gallon = _____ quarts

Temperature

15. Water freezes at _____ °F.

16. Water boils at _____ °F.

17. Normal body temperature is _____ °F.

Customary to Metric

18. 1 inch = _____ centimeters

Metric Units

Linear Measure

19. 1 centimeter = _____ millimeters

20. 1 meter = _____ centimeters

21. 1 meter = _____ millimeters

22. 1 kilometer = _____ meters

Area

23. 1 meter^2 = _____ centimeters^2

24. 1 kilometer^2 = _____ meters^2

Volume

25. 1 meter^3 = _____ centimeters^3

Mass

26. 1 gram = _____ milligrams

27. 1 kilogram = _____ grams

28. 1 metric ton = _____ kilograms

Capacity

29. 1 liter = _____ milliliters

30. One milliliter of water has a volume of _____ and a mass of _____. One thousand cm^3 of water fills a _____-liter container and has a mass of _____ kilogram.

Temperature

31. Water freezes at _____ °C.

32. Water boils at _____ °C.

33. Normal body temperature is _____ °C.

G $+ - \times \div$ **Fractions**
For use with Lesson 38

Name _____

Time _____

Simplify these expressions. Reduce the answers.

$\dfrac{2}{3} + \dfrac{2}{3} =$	$\dfrac{2}{3} - \dfrac{2}{3} =$	$\dfrac{2}{3} \times \dfrac{2}{3} =$	$\dfrac{2}{3} \div \dfrac{2}{3} =$
$\dfrac{3}{4} + \dfrac{1}{4} =$	$\dfrac{3}{4} - \dfrac{1}{4} =$	$\dfrac{3}{4} \times \dfrac{1}{4} =$	$\dfrac{3}{4} \div \dfrac{1}{4} =$
$\dfrac{2}{3} + \dfrac{1}{2} =$	$\dfrac{2}{3} - \dfrac{1}{2} =$	$\dfrac{2}{3} \times \dfrac{1}{2} =$	$\dfrac{2}{3} \div \dfrac{1}{2} =$
$\dfrac{3}{4} + \dfrac{2}{3} =$	$\dfrac{3}{4} - \dfrac{2}{3} =$	$\dfrac{3}{4} \times \dfrac{2}{3} =$	$\dfrac{3}{4} \div \dfrac{2}{3} =$
$\dfrac{2}{5} + \dfrac{1}{4} =$	$\dfrac{2}{5} - \dfrac{1}{4} =$	$\dfrac{2}{5} \times \dfrac{1}{4} =$	$\dfrac{2}{5} \div \dfrac{1}{4} =$
$\dfrac{1}{2} + \dfrac{5}{8} =$	$\dfrac{5}{8} - \dfrac{1}{2} =$	$\dfrac{1}{2} \times \dfrac{5}{8} =$	$\dfrac{1}{2} \div \dfrac{5}{8} =$

H

Measurement Facts
For use with Lesson 39

Name _____

Time _____

Write the number that makes each statement true.

Customary Units

Linear Measure

1. 1 foot = _____ inches

2. 1 yard = _____ inches

3. 1 yard = _____ feet

4. 1 mile = _____ feet

5. 1 mile = _____ yards

Area

6. 1 foot^2 = _____ inches^2

7. 1 yard^2 = _____ feet^2

Volume

8. 1 yard^3 = _____ feet^3

Weight

9. 1 pound = _____ ounces

10. 1 ton = _____ pounds

Liquid Measure

11. 1 pint = _____ ounces

12. 1 pint = _____ cups

13. 1 quart = _____ pints

14. 1 gallon = _____ quarts

Temperature

15. Water freezes at _____ °F.

16. Water boils at _____ °F.

17. Normal body temperature is _____ °F.

Customary to Metric

18. 1 inch = _____ centimeters

Metric Units

Linear Measure

19. 1 centimeter = _____ millimeters

20. 1 meter = _____ centimeters

21. 1 meter = _____ millimeters

22. 1 kilometer = _____ meters

Area

23. 1 meter^2 = _____ centimeters^2

24. 1 kilometer^2 = _____ meters^2

Volume

25. 1 meter^3 = _____ centimeters^3

Mass

26. 1 gram = _____ milligrams

27. 1 kilogram = _____ grams

28. 1 metric ton = _____ kilograms

Capacity

29. 1 liter = _____ milliliters

30. One milliliter of water has a volume of _____ and a mass of _____. One thousand cm^3 of water fills a _____-liter container and has a mass of _____ kilogram.

Temperature

31. Water freezes at _____ °C.

32. Water boils at _____ °C.

33. Normal body temperature is _____ °C.

G   **+ − × ÷ Fractions**
For use with Lesson 40

Name _____

Time _____

Simplify these expressions. Reduce the answers.

$\frac{2}{3} + \frac{2}{3} =$	$\frac{2}{3} - \frac{2}{3} =$	$\frac{2}{3} \times \frac{2}{3} =$	$\frac{2}{3} \div \frac{2}{3} =$
$\frac{3}{4} + \frac{1}{4} =$	$\frac{3}{4} - \frac{1}{4} =$	$\frac{3}{4} \times \frac{1}{4} =$	$\frac{3}{4} \div \frac{1}{4} =$
$\frac{2}{3} + \frac{1}{2} =$	$\frac{2}{3} - \frac{1}{2} =$	$\frac{2}{3} \times \frac{1}{2} =$	$\frac{2}{3} \div \frac{1}{2} =$
$\frac{3}{4} + \frac{2}{3} =$	$\frac{3}{4} - \frac{2}{3} =$	$\frac{3}{4} \times \frac{2}{3} =$	$\frac{3}{4} \div \frac{2}{3} =$
$\frac{2}{5} + \frac{1}{4} =$	$\frac{2}{5} - \frac{1}{4} =$	$\frac{2}{5} \times \frac{1}{4} =$	$\frac{2}{5} \div \frac{1}{4} =$
$\frac{1}{2} + \frac{5}{8} =$	$\frac{5}{8} - \frac{1}{2} =$	$\frac{1}{2} \times \frac{5}{8} =$	$\frac{1}{2} \div \frac{5}{8} =$

G   + − × ÷ **Fractions**
For use with Test 7

Name _____

Time _____

Simplify these expressions. Reduce the answers.

$\frac{2}{3} + \frac{2}{3} =$	$\frac{2}{3} - \frac{2}{3} =$	$\frac{2}{3} \times \frac{2}{3} =$	$\frac{2}{3} \div \frac{2}{3} =$
$\frac{3}{4} + \frac{1}{4} =$	$\frac{3}{4} - \frac{1}{4} =$	$\frac{3}{4} \times \frac{1}{4} =$	$\frac{3}{4} \div \frac{1}{4} =$
$\frac{2}{3} + \frac{1}{2} =$	$\frac{2}{3} - \frac{1}{2} =$	$\frac{2}{3} \times \frac{1}{2} =$	$\frac{2}{3} \div \frac{1}{2} =$
$\frac{3}{4} + \frac{2}{3} =$	$\frac{3}{4} - \frac{2}{3} =$	$\frac{3}{4} \times \frac{2}{3} =$	$\frac{3}{4} \div \frac{2}{3} =$
$\frac{2}{5} + \frac{1}{4} =$	$\frac{2}{5} - \frac{1}{4} =$	$\frac{2}{5} \times \frac{1}{4} =$	$\frac{2}{5} \div \frac{1}{4} =$
$\frac{1}{2} + \frac{5}{8} =$	$\frac{5}{8} - \frac{1}{2} =$	$\frac{1}{2} \times \frac{5}{8} =$	$\frac{1}{2} \div \frac{5}{8} =$

I

Proportions
For use with Lesson 41

Name _____

Time _____

Find the number that completes each proportion.

$\dfrac{3}{4} = \dfrac{a}{12}$ $a =$	$\dfrac{3}{4} = \dfrac{12}{b}$ $b =$	$\dfrac{c}{5} = \dfrac{12}{20}$ $c =$	$\dfrac{2}{d} = \dfrac{12}{24}$ $d =$
$\dfrac{4}{10} = \dfrac{e}{30}$ $e =$	$\dfrac{8}{12} = \dfrac{4}{f}$ $f =$	$\dfrac{g}{10} = \dfrac{10}{5}$ $g =$	$\dfrac{5}{h} = \dfrac{6}{18}$ $h =$
$\dfrac{15}{20} = \dfrac{i}{40}$ $i =$	$\dfrac{25}{100} = \dfrac{5}{j}$ $j =$	$\dfrac{k}{30} = \dfrac{3}{9}$ $k =$	$\dfrac{5}{m} = \dfrac{10}{100}$ $m =$
$\dfrac{50}{100} = \dfrac{n}{30}$ $n =$	$\dfrac{20}{15} = \dfrac{60}{p}$ $p =$	$\dfrac{q}{40} = \dfrac{75}{100}$ $q =$	$\dfrac{5}{r} = \dfrac{4}{16}$ $r =$
$\dfrac{2}{5} = \dfrac{s}{100}$ $s =$	$\dfrac{6}{8} = \dfrac{9}{t}$ $t =$	$\dfrac{u}{16} = \dfrac{8}{4}$ $u =$	$\dfrac{60}{v} = \dfrac{3}{2}$ $v =$
$\dfrac{8}{10} = \dfrac{w}{100}$ $w =$	$\dfrac{9}{12} = \dfrac{36}{x}$ $x =$	$\dfrac{y}{30} = \dfrac{6}{20}$ $y =$	$\dfrac{24}{z} = \dfrac{8}{6}$ $z =$

H | Measurement Facts

For use with Lesson 42

Name _____

Time _____

Write the number that makes each statement true.

Customary Units

Linear Measure

1. 1 foot = _____ inches

2. 1 yard = _____ inches

3. 1 yard = _____ feet

4. 1 mile = _____ feet

5. 1 mile = _____ yards

Area

6. 1 foot^2 = _____ inches^2

7. 1 yard^2 = _____ feet^2

Volume

8. 1 yard^3 = _____ feet^3

Weight

9. 1 pound = _____ ounces

10. 1 ton = _____ pounds

Liquid Measure

11. 1 pint = _____ ounces

12. 1 pint = _____ cups

13. 1 quart = _____ pints

14. 1 gallon = _____ quarts

Temperature

15. Water freezes at _____ °F.

16. Water boils at _____ °F.

17. Normal body temperature is _____ °F.

Customary to Metric

18. 1 inch = _____ centimeters

Metric Units

Linear Measure

19. 1 centimeter = _____ millimeters

20. 1 meter = _____ centimeters

21. 1 meter = _____ millimeters

22. 1 kilometer = _____ meters

Area

23. 1 meter^2 = _____ centimeters^2

24. 1 kilometer^2 = _____ meters^2

Volume

25. 1 meter^3 = _____ centimeters^3

Mass

26. 1 gram = _____ milligrams

27. 1 kilogram = _____ grams

28. 1 metric ton = _____ kilograms

Capacity

29. 1 liter = _____ milliliters

30. One milliliter of water has a volume of _____ and a mass of _____. One thousand cm^3 of water fills a _____-liter container and has a mass of _____ kilogram.

Temperature

31. Water freezes at _____ °C.

32. Water boils at _____ °C.

33. Normal body temperature is _____ °C.

I

Proportions
For use with Lesson 43

Name _____

Time _____

Find the number that completes each proportion.

$\dfrac{3}{4} = \dfrac{a}{12}$ $a =$	$\dfrac{3}{4} = \dfrac{12}{b}$ $b =$	$\dfrac{c}{5} = \dfrac{12}{20}$ $c =$	$\dfrac{2}{d} = \dfrac{12}{24}$ $d =$
$\dfrac{4}{10} = \dfrac{e}{30}$ $e =$	$\dfrac{8}{12} = \dfrac{4}{f}$ $f =$	$\dfrac{g}{10} = \dfrac{10}{5}$ $g =$	$\dfrac{5}{h} = \dfrac{6}{18}$ $h =$
$\dfrac{15}{20} = \dfrac{i}{40}$ $i =$	$\dfrac{25}{100} = \dfrac{5}{j}$ $j =$	$\dfrac{k}{30} = \dfrac{3}{9}$ $k =$	$\dfrac{5}{m} = \dfrac{10}{100}$ $m =$
$\dfrac{50}{100} = \dfrac{n}{30}$ $n =$	$\dfrac{20}{15} = \dfrac{60}{p}$ $p =$	$\dfrac{q}{40} = \dfrac{75}{100}$ $q =$	$\dfrac{5}{r} = \dfrac{4}{16}$ $r =$
$\dfrac{2}{5} = \dfrac{s}{100}$ $s =$	$\dfrac{6}{8} = \dfrac{9}{t}$ $t =$	$\dfrac{u}{16} = \dfrac{8}{4}$ $u =$	$\dfrac{60}{v} = \dfrac{3}{2}$ $v =$
$\dfrac{8}{10} = \dfrac{w}{100}$ $w =$	$\dfrac{9}{12} = \dfrac{36}{x}$ $x =$	$\dfrac{y}{30} = \dfrac{6}{20}$ $y =$	$\dfrac{24}{z} = \dfrac{8}{6}$ $z =$

Measurement Facts
For use with Lesson 44

Name _____

Time _____

Write the number that makes each statement true.

Customary Units	Metric Units

Linear Measure

1. 1 foot = _____ inches

2. 1 yard = _____ inches

3. 1 yard = _____ feet

4. 1 mile = _____ feet

5. 1 mile = _____ yards

Area

6. 1 foot^2 = _____ inches^2

7. 1 yard^2 = _____ feet^2

Volume

8. 1 yard^3 = _____ feet^3

Weight

9. 1 pound = _____ ounces

10. 1 ton = _____ pounds

Liquid Measure

11. 1 pint = _____ ounces

12. 1 pint = _____ cups

13. 1 quart = _____ pints

14. 1 gallon = _____ quarts

Temperature

15. Water freezes at _____ °F.

16. Water boils at _____ °F.

17. Normal body temperature is _____ °F.

Customary to Metric

18. 1 inch = _____ centimeters

Linear Measure

19. 1 centimeter = _____ millimeters

20. 1 meter = _____ centimeters

21. 1 meter = _____ millimeters

22. 1 kilometer = _____ meters

Area

23. 1 meter^2 = _____ centimeters^2

24. 1 kilometer^2 = _____ meters^2

Volume

25. 1 meter^3 = _____ centimeters^3

Mass

26. 1 gram = _____ milligrams

27. 1 kilogram = _____ grams

28. 1 metric ton = _____ kilograms

Capacity

29. 1 liter = _____ milliliters

30. One milliliter of water has a volume of _____ and a mass of _____. One thousand cm^3 of water fills a _____-liter container and has a mass of _____ kilogram.

Temperature

31. Water freezes at _____ °C.

32. Water boils at _____ °C.

33. Normal body temperature is _____ °C.

Saxon Math 8/7—Homeschool

I Proportions
For use with Lesson 45

Name _____

Time _____

Find the number that completes each proportion.

$\frac{3}{4} = \frac{a}{12}$ $a =$	$\frac{3}{4} = \frac{12}{b}$ $b =$	$\frac{c}{5} = \frac{12}{20}$ $c =$	$\frac{2}{d} = \frac{12}{24}$ $d =$
$\frac{4}{10} = \frac{e}{30}$ $e =$	$\frac{8}{12} = \frac{4}{f}$ $f =$	$\frac{g}{10} = \frac{10}{5}$ $g =$	$\frac{5}{h} = \frac{6}{18}$ $h =$
$\frac{15}{20} = \frac{i}{40}$ $i =$	$\frac{25}{100} = \frac{5}{j}$ $j =$	$\frac{k}{30} = \frac{3}{9}$ $k =$	$\frac{5}{m} = \frac{10}{100}$ $m =$
$\frac{50}{100} = \frac{n}{30}$ $n =$	$\frac{20}{15} = \frac{60}{p}$ $p =$	$\frac{q}{40} = \frac{75}{100}$ $q =$	$\frac{5}{r} = \frac{4}{16}$ $r =$
$\frac{2}{5} = \frac{s}{100}$ $s =$	$\frac{6}{8} = \frac{9}{t}$ $t =$	$\frac{u}{16} = \frac{8}{4}$ $u =$	$\frac{60}{v} = \frac{3}{2}$ $v =$
$\frac{8}{10} = \frac{w}{100}$ $w =$	$\frac{9}{12} = \frac{36}{x}$ $x =$	$\frac{y}{30} = \frac{6}{20}$ $y =$	$\frac{24}{z} = \frac{8}{6}$ $z =$

H

Measurement Facts
For use with Test 8

Name _____

Time _____

Write the number that makes each statement true.

Customary Units

Linear Measure

1. 1 foot = _____ inches

2. 1 yard = _____ inches

3. 1 yard = _____ feet

4. 1 mile = _____ feet

5. 1 mile = _____ yards

Area

6. 1 foot^2 = _____ inches^2

7. 1 yard^2 = _____ feet^2

Volume

8. 1 yard^3 = _____ feet^3

Weight

9. 1 pound = _____ ounces

10. 1 ton = _____ pounds

Liquid Measure

11. 1 pint = _____ ounces

12. 1 pint = _____ cups

13. 1 quart = _____ pints

14. 1 gallon = _____ quarts

Temperature

15. Water freezes at _____ °F.

16. Water boils at _____ °F.

17. Normal body temperature is _____ °F.

Customary to Metric

18. 1 inch = _____ centimeters

Metric Units

Linear Measure

19. 1 centimeter = _____ millimeters

20. 1 meter = _____ centimeters

21. 1 meter = _____ millimeters

22. 1 kilometer = _____ meters

Area

23. 1 meter^2 = _____ centimeters^2

24. 1 kilometer^2 = _____ meters^2

Volume

25. 1 meter^3 = _____ centimeters^3

Mass

26. 1 gram = _____ milligrams

27. 1 kilogram = _____ grams

28. 1 metric ton = _____ kilograms

Capacity

29. 1 liter = _____ milliliters

30. One milliliter of water has a volume of _____ and a mass of _____. One thousand cm^3 of water fills a _____-liter container and has a mass of _____ kilogram.

Temperature

31. Water freezes at _____ °C.

32. Water boils at _____ °C.

33. Normal body temperature is _____ °C.

J

+ − × ÷ Decimals
For use with Lesson 46

Name _____

Time _____

Simplify these expressions.

0.8 + 0.4 =	0.8 × 0.4 =	0.8 ÷ 0.4 =
1.2 − 0.4 =	1.2 × 0.4 =	1.2 ÷ 0.4 =
1.2 + 0.04 =	1.2 × 0.04 =	1.2 ÷ 0.04 =
1.2 + 4 =	1.2 × 4 =	1.2 ÷ 4 =
6 − 0.3 =	6 × 0.3 =	6 ÷ 0.3 =
0.3 + 6 =	0.3 × 6 =	0.3 ÷ 6 =
0.01 − 0.01 =	0.01 × 0.01 =	0.01 ÷ 0.01 =

I Proportions
For use with Lesson 47

Name _____

Time _____

Find the number that completes each proportion.

$\dfrac{3}{4} = \dfrac{a}{12}$ $a =$	$\dfrac{3}{4} = \dfrac{12}{b}$ $b =$	$\dfrac{c}{5} = \dfrac{12}{20}$ $c =$	$\dfrac{2}{d} = \dfrac{12}{24}$ $d =$
$\dfrac{4}{10} = \dfrac{e}{30}$ $e =$	$\dfrac{8}{12} = \dfrac{4}{f}$ $f =$	$\dfrac{g}{10} = \dfrac{10}{5}$ $g =$	$\dfrac{5}{h} = \dfrac{6}{18}$ $h =$
$\dfrac{15}{20} = \dfrac{i}{40}$ $i =$	$\dfrac{25}{100} = \dfrac{5}{j}$ $j =$	$\dfrac{k}{30} = \dfrac{3}{9}$ $k =$	$\dfrac{5}{m} = \dfrac{10}{100}$ $m =$
$\dfrac{50}{100} = \dfrac{n}{30}$ $n =$	$\dfrac{20}{15} = \dfrac{60}{p}$ $p =$	$\dfrac{q}{40} = \dfrac{75}{100}$ $q =$	$\dfrac{5}{r} = \dfrac{4}{16}$ $r =$
$\dfrac{2}{5} = \dfrac{s}{100}$ $s =$	$\dfrac{6}{8} = \dfrac{9}{t}$ $t =$	$\dfrac{u}{16} = \dfrac{8}{4}$ $u =$	$\dfrac{60}{v} = \dfrac{3}{2}$ $v =$
$\dfrac{8}{10} = \dfrac{w}{100}$ $w =$	$\dfrac{9}{12} = \dfrac{36}{x}$ $x =$	$\dfrac{y}{30} = \dfrac{6}{20}$ $y =$	$\dfrac{24}{z} = \dfrac{8}{6}$ $z =$

J

+ − × ÷ **Decimals**
For use with Lesson 48

Name _____

Time _____

Simplify these expressions.

0.8 + 0.4 =	0.8 × 0.4 =	0.8 ÷ 0.4 =
1.2 − 0.4 =	1.2 × 0.4 =	1.2 ÷ 0.4 =
1.2 + 0.04 =	1.2 × 0.04 =	1.2 ÷ 0.04 =
1.2 + 4 =	1.2 × 4 =	1.2 ÷ 4 =
6 − 0.3 =	6 × 0.3 =	6 ÷ 0.3 =
0.3 + 6 =	0.3 × 6 =	0.3 ÷ 6 =
0.01 − 0.01 =	0.01 × 0.01 =	0.01 ÷ 0.01 =

J

+ − × ÷ Decimals
For use with Lesson 49

Name _____

Time _____

Simplify these expressions.

0.8 + 0.4 =	0.8 × 0.4 =	0.8 ÷ 0.4 =
1.2 − 0.4 =	1.2 × 0.4 =	1.2 ÷ 0.4 =
1.2 + 0.04 =	1.2 × 0.04 =	1.2 ÷ 0.04 =
1.2 + 4 =	1.2 × 4 =	1.2 ÷ 4 =
6 − 0.3 =	6 × 0.3 =	6 ÷ 0.3 =
0.3 + 6 =	0.3 × 6 =	0.3 ÷ 6 =
0.01 − 0.01 =	0.01 × 0.01 =	0.01 ÷ 0.01 =

Saxon Math 8/7—Homeschool

I	**Proportions**	Name _____
	For use with Lesson 50	Time _____

Find the number that completes each proportion.

$\dfrac{3}{4} = \dfrac{a}{12}$ $a =$	$\dfrac{3}{4} = \dfrac{12}{b}$ $b =$	$\dfrac{c}{5} = \dfrac{12}{20}$ $c =$	$\dfrac{2}{d} = \dfrac{12}{24}$ $d =$
$\dfrac{4}{10} = \dfrac{e}{30}$ $e =$	$\dfrac{8}{12} = \dfrac{4}{f}$ $f =$	$\dfrac{g}{10} = \dfrac{10}{5}$ $g =$	$\dfrac{5}{h} = \dfrac{6}{18}$ $h =$
$\dfrac{15}{20} = \dfrac{i}{40}$ $i =$	$\dfrac{25}{100} = \dfrac{5}{j}$ $j =$	$\dfrac{k}{30} = \dfrac{3}{9}$ $k =$	$\dfrac{5}{m} = \dfrac{10}{100}$ $m =$
$\dfrac{50}{100} = \dfrac{n}{30}$ $n =$	$\dfrac{20}{15} = \dfrac{60}{p}$ $p =$	$\dfrac{q}{40} = \dfrac{75}{100}$ $q =$	$\dfrac{5}{r} = \dfrac{4}{16}$ $r =$
$\dfrac{2}{5} = \dfrac{s}{100}$ $s =$	$\dfrac{6}{8} = \dfrac{9}{t}$ $t =$	$\dfrac{u}{16} = \dfrac{8}{4}$ $u =$	$\dfrac{60}{v} = \dfrac{3}{2}$ $v =$
$\dfrac{8}{10} = \dfrac{w}{100}$ $w =$	$\dfrac{9}{12} = \dfrac{36}{x}$ $x =$	$\dfrac{y}{30} = \dfrac{6}{20}$ $y =$	$\dfrac{24}{z} = \dfrac{8}{6}$ $z =$

I | Proportions
For use with Test 9

Name _____

Time _____

Find the number that completes each proportion.

$\dfrac{3}{4} = \dfrac{a}{12}$ $a =$	$\dfrac{3}{4} = \dfrac{12}{b}$ $b =$	$\dfrac{c}{5} = \dfrac{12}{20}$ $c =$	$\dfrac{2}{d} = \dfrac{12}{24}$ $d =$
$\dfrac{4}{10} = \dfrac{e}{30}$ $e =$	$\dfrac{8}{12} = \dfrac{4}{f}$ $f =$	$\dfrac{g}{10} = \dfrac{10}{5}$ $g =$	$\dfrac{5}{h} = \dfrac{6}{18}$ $h =$
$\dfrac{15}{20} = \dfrac{i}{40}$ $i =$	$\dfrac{25}{100} = \dfrac{5}{j}$ $j =$	$\dfrac{k}{30} = \dfrac{3}{9}$ $k =$	$\dfrac{5}{m} = \dfrac{10}{100}$ $m =$
$\dfrac{50}{100} = \dfrac{n}{30}$ $n =$	$\dfrac{20}{15} = \dfrac{60}{p}$ $p =$	$\dfrac{q}{40} = \dfrac{75}{100}$ $q =$	$\dfrac{5}{r} = \dfrac{4}{16}$ $r =$
$\dfrac{2}{5} = \dfrac{s}{100}$ $s =$	$\dfrac{6}{8} = \dfrac{9}{t}$ $t =$	$\dfrac{u}{16} = \dfrac{8}{4}$ $u =$	$\dfrac{60}{v} = \dfrac{3}{2}$ $v =$
$\dfrac{8}{10} = \dfrac{w}{100}$ $w =$	$\dfrac{9}{12} = \dfrac{36}{x}$ $x =$	$\dfrac{y}{30} = \dfrac{6}{20}$ $y =$	$\dfrac{24}{z} = \dfrac{8}{6}$ $z =$

J

+ − × ÷ **Decimals**
For use with Lesson 51

Name _____

Time _____

Simplify these expressions.

0.8 + 0.4 =	0.8 × 0.4 =	0.8 ÷ 0.4 =
1.2 − 0.4 =	1.2 × 0.4 =	1.2 ÷ 0.4 =
1.2 + 0.04 =	1.2 × 0.04 =	1.2 ÷ 0.04 =
1.2 + 4 =	1.2 × 4 =	1.2 ÷ 4 =
6 − 0.3 =	6 × 0.3 =	6 ÷ 0.3 =
0.3 + 6 =	0.3 × 6 =	0.3 ÷ 6 =
0.01 − 0.01 =	0.01 × 0.01 =	0.01 ÷ 0.01 =

K

Powers and Roots
For use with Lesson 52

Name _____

Time _____

Simplify each power or root.

$\sqrt{100} =$	$\sqrt{16} =$	$\sqrt{81} =$	$\sqrt{4} =$
$\sqrt{144} =$	$\sqrt{1} =$	$\sqrt{64} =$	$\sqrt{49} =$
$\sqrt{25} =$	$\sqrt{121} =$	$\sqrt{9} =$	$\sqrt{36} =$
$\sqrt{169} =$	$\sqrt{225} =$	$\sqrt{196} =$	$\sqrt{625} =$
$8^2 =$	$5^2 =$	$3^2 =$	$12^2 =$
$10^2 =$	$2^3 =$	$6^2 =$	$3^3 =$
$4^2 =$	$10^3 =$	$7^2 =$	$15^2 =$
$5^3 =$	$25^2 =$	$4^3 =$	$9^2 =$

Powers and Roots
For use with Lesson 53

Name _____

Time _____

Simplify each power or root.

$\sqrt{100} =$	$\sqrt{16} =$	$\sqrt{81} =$	$\sqrt{4} =$
$\sqrt{144} =$	$\sqrt{1} =$	$\sqrt{64} =$	$\sqrt{49} =$
$\sqrt{25} =$	$\sqrt{121} =$	$\sqrt{9} =$	$\sqrt{36} =$
$\sqrt{169} =$	$\sqrt{225} =$	$\sqrt{196} =$	$\sqrt{625} =$
$8^2 =$	$5^2 =$	$3^2 =$	$12^2 =$
$10^2 =$	$2^3 =$	$6^2 =$	$3^3 =$
$4^2 =$	$10^3 =$	$7^2 =$	$15^2 =$
$5^3 =$	$25^2 =$	$4^3 =$	$9^2 =$

L

Fraction-Decimal-Percent Equivalents
For use with Lesson 54

Name _____

Time _____

Write each fraction as a decimal and as a percent. Write repeating decimals with a bar over the repetend.

Fraction	Decimal	Percent
$\frac{1}{2}$		
$\frac{1}{3}$		
$\frac{2}{3}$		
$\frac{1}{4}$		
$\frac{3}{4}$		
$\frac{1}{5}$		
$\frac{2}{5}$		
$\frac{3}{5}$		
$\frac{4}{5}$		
$\frac{1}{6}$		
$\frac{5}{6}$		
$\frac{1}{8}$		
$\frac{3}{8}$		

Fraction	Decimal	Percent
$\frac{5}{8}$		
$\frac{7}{8}$		
$\frac{1}{9}$		
$\frac{1}{10}$		
$\frac{3}{10}$		
$\frac{7}{10}$		
$\frac{9}{10}$		
$\frac{1}{20}$		
$\frac{1}{25}$		
$\frac{1}{50}$		
$\frac{1}{100}$		
$1\frac{1}{2}$		

J **+ − × ÷ Decimals**
For use with Lesson 55

Name _____

Time _____

Simplify these expressions.

0.8 + 0.4 =	0.8 × 0.4 =	0.8 ÷ 0.4 =
1.2 − 0.4 =	1.2 × 0.4 =	1.2 ÷ 0.4 =
1.2 + 0.04 =	1.2 × 0.04 =	1.2 ÷ 0.04 =
1.2 + 4 =	1.2 × 4 =	1.2 ÷ 4 =
6 − 0.3 =	6 × 0.3 =	6 ÷ 0.3 =
0.3 + 6 =	0.3 × 6 =	0.3 ÷ 6 =
0.01 − 0.01 =	0.01 × 0.01 =	0.01 ÷ 0.01 =

J   + − × ÷ **Decimals**
For use with Test 10

Name _____

Time _____

Simplify these expressions.

0.8 + 0.4 =	0.8 × 0.4 =	0.8 ÷ 0.4 =
1.2 − 0.4 =	1.2 × 0.4 =	1.2 ÷ 0.4 =
1.2 + 0.04 =	1.2 × 0.04 =	1.2 ÷ 0.04 =
1.2 + 4 =	1.2 × 4 =	1.2 ÷ 4 =
6 − 0.3 =	6 × 0.3 =	6 ÷ 0.3 =
0.3 + 6 =	0.3 × 6 =	0.3 ÷ 6 =
0.01 − 0.01 =	0.01 × 0.01 =	0.01 ÷ 0.01 =

L

Fraction-Decimal-Percent Equivalents

For use with Lesson 56

Name _____

Time _____

Write each fraction as a decimal and as a percent. Write repeating decimals with a bar over the repetend.

Fraction	Decimal	Percent	Fraction	Decimal	Percent
$\frac{1}{2}$			$\frac{5}{8}$		
$\frac{1}{3}$			$\frac{7}{8}$		
$\frac{2}{3}$			$\frac{1}{9}$		
$\frac{1}{4}$			$\frac{1}{10}$		
$\frac{3}{4}$			$\frac{3}{10}$		
$\frac{1}{5}$			$\frac{7}{10}$		
$\frac{2}{5}$			$\frac{9}{10}$		
$\frac{3}{5}$			$\frac{1}{20}$		
$\frac{4}{5}$			$\frac{1}{25}$		
$\frac{1}{6}$			$\frac{1}{50}$		
$\frac{5}{6}$			$\frac{1}{100}$		
$\frac{1}{8}$			$1\frac{1}{2}$		
$\frac{3}{8}$					

K

Powers and Roots
For use with Lesson 57

Name _____

Time _____

Simplify each power or root.

$\sqrt{100} =$	$\sqrt{16} =$	$\sqrt{81} =$	$\sqrt{4} =$
$\sqrt{144} =$	$\sqrt{1} =$	$\sqrt{64} =$	$\sqrt{49} =$
$\sqrt{25} =$	$\sqrt{121} =$	$\sqrt{9} =$	$\sqrt{36} =$
$\sqrt{169} =$	$\sqrt{225} =$	$\sqrt{196} =$	$\sqrt{625} =$
$8^2 =$	$5^2 =$	$3^2 =$	$12^2 =$
$10^2 =$	$2^3 =$	$6^2 =$	$3^3 =$
$4^2 =$	$10^3 =$	$7^2 =$	$15^2 =$
$5^3 =$	$25^2 =$	$4^3 =$	$9^2 =$

Powers and Roots
For use with Lesson 58

Name _____

Time _____

Simplify each power or root.

$\sqrt{100} =$	$\sqrt{16} =$	$\sqrt{81} =$	$\sqrt{4} =$
$\sqrt{144} =$	$\sqrt{1} =$	$\sqrt{64} =$	$\sqrt{49} =$
$\sqrt{25} =$	$\sqrt{121} =$	$\sqrt{9} =$	$\sqrt{36} =$
$\sqrt{169} =$	$\sqrt{225} =$	$\sqrt{196} =$	$\sqrt{625} =$
$8^2 =$	$5^2 =$	$3^2 =$	$12^2 =$
$10^2 =$	$2^3 =$	$6^2 =$	$3^3 =$
$4^2 =$	$10^3 =$	$7^2 =$	$15^2 =$
$5^3 =$	$25^2 =$	$4^3 =$	$9^2 =$

L	**Fraction-Decimal-Percent Equivalents**

For use with Lesson 59

Name _____

Time _____

Write each fraction as a decimal and as a percent. Write repeating decimals with a bar over the repetend.

Fraction	Decimal	Percent	Fraction	Decimal	Percent
$\frac{1}{2}$			$\frac{5}{8}$		
$\frac{1}{3}$			$\frac{7}{8}$		
$\frac{2}{3}$			$\frac{1}{9}$		
$\frac{1}{4}$			$\frac{1}{10}$		
$\frac{3}{4}$			$\frac{3}{10}$		
$\frac{1}{5}$			$\frac{7}{10}$		
$\frac{2}{5}$			$\frac{9}{10}$		
$\frac{3}{5}$			$\frac{1}{20}$		
$\frac{4}{5}$			$\frac{1}{25}$		
$\frac{1}{6}$			$\frac{1}{50}$		
$\frac{5}{6}$			$\frac{1}{100}$		
$\frac{1}{8}$			$1\frac{1}{2}$		
$\frac{3}{8}$					

M	**Metric Conversions** *For use with Lesson 60*

Name _____

Time _____

Complete each equivalence.

1. 2 meters = _____ centimeters

2. 1.5 kilometers = _____ meters

3. 2.54 centimeters = _____ millimeters

4. 125 centimeters = _____ meters

5. 75 millimeters = _____ centimeters

6. 0.8 meter = _____ millimeters

7. 10 kilometers = _____ meters

8. 0.1 kilometer = _____ meters

9. 5000 meters = _____ kilometers

10. 50 centimeters = _____ meter

11. 50 centimeters = _____ millimeters

12. 2 liters = _____ milliliters

13. 250 milliliters = _____ liter

14. 4 kilograms = _____ grams

15. 2.5 grams = _____ milligrams

16. 500 milligrams = _____ gram

17. 0.5 kilogram = _____ grams

18. Two liters of water has a volume of

_____ cubic centimeters and a mass

of _____ kilograms.

Record the factor indicated by each prefix.

	Prefix	Factor
19.	kilo-	
20.	hecto-	
21.	deka-	
	(unit)	1
22.	deci-	
23.	centi-	
24.	milli-	

K | Powers and Roots
For use with Test 11

Name _____

Time _____

Simplify each power or root.

$\sqrt{100} =$	$\sqrt{16} =$	$\sqrt{81} =$	$\sqrt{4} =$
$\sqrt{144} =$	$\sqrt{1} =$	$\sqrt{64} =$	$\sqrt{49} =$
$\sqrt{25} =$	$\sqrt{121} =$	$\sqrt{9} =$	$\sqrt{36} =$
$\sqrt{169} =$	$\sqrt{225} =$	$\sqrt{196} =$	$\sqrt{625} =$
$8^2 =$	$5^2 =$	$3^2 =$	$12^2 =$
$10^2 =$	$2^3 =$	$6^2 =$	$3^3 =$
$4^2 =$	$10^3 =$	$7^2 =$	$15^2 =$
$5^3 =$	$25^2 =$	$4^3 =$	$9^2 =$

Saxon Math 8/7—Homeschool

Fraction-Decimal-Percent Equivalents

L

For use with Lesson 61

Name _____

Time _____

Write each fraction as a decimal and as a percent. Write repeating decimals with a bar over the repetend.

Fraction	Decimal	Percent	Fraction	Decimal	Percent
$\frac{1}{2}$			$\frac{5}{8}$		
$\frac{1}{3}$			$\frac{7}{8}$		
$\frac{2}{3}$			$\frac{1}{9}$		
$\frac{1}{4}$			$\frac{1}{10}$		
$\frac{3}{4}$			$\frac{3}{10}$		
$\frac{1}{5}$			$\frac{7}{10}$		
$\frac{2}{5}$			$\frac{9}{10}$		
$\frac{3}{5}$			$\frac{1}{20}$		
$\frac{4}{5}$			$\frac{1}{25}$		
$\frac{1}{6}$			$\frac{1}{50}$		
$\frac{5}{6}$			$\frac{1}{100}$		
$\frac{1}{8}$			$1\frac{1}{2}$		
$\frac{3}{8}$					

M | Metric Conversions
For use with Lesson 62

Name _____

Time _____

Complete each equivalence.

1. 2 meters = _____ centimeters

2. 1.5 kilometers = _____ meters

3. 2.54 centimeters = _____ millimeters

4. 125 centimeters = _____ meters

5. 75 millimeters = _____ centimeters

6. 0.8 meter = _____ millimeters

7. 10 kilometers = _____ meters

8. 0.1 kilometer = _____ meters

9. 5000 meters = _____ kilometers

10. 50 centimeters = _____ meter

11. 50 centimeters = _____ millimeters

12. 2 liters = _____ milliliters

13. 250 milliliters = _____ liter

14. 4 kilograms = _____ grams

15. 2.5 grams = _____ milligrams

16. 500 milligrams = _____ gram

17. 0.5 kilogram = _____ grams

18. Two liters of water has a volume of

_____ cubic centimeters and a mass

of _____ kilograms.

Record the factor indicated by each prefix.

	Prefix	Factor
19.	kilo-	
20.	hecto-	
21.	deka-	
	(unit)	1
22.	deci-	
23.	centi-	
24.	milli-	

L — Fraction-Decimal-Percent Equivalents
For use with Lesson 63

Name _____

Time _____

Write each fraction as a decimal and as a percent. Write repeating decimals with a bar over the repetend.

Fraction	Decimal	Percent	Fraction	Decimal	Percent
$\frac{1}{2}$			$\frac{5}{8}$		
$\frac{1}{3}$			$\frac{7}{8}$		
$\frac{2}{3}$			$\frac{1}{9}$		
$\frac{1}{4}$			$\frac{1}{10}$		
$\frac{3}{4}$			$\frac{3}{10}$		
$\frac{1}{5}$			$\frac{7}{10}$		
$\frac{2}{5}$			$\frac{9}{10}$		
$\frac{3}{5}$			$\frac{1}{20}$		
$\frac{4}{5}$			$\frac{1}{25}$		
$\frac{1}{6}$			$\frac{1}{50}$		
$\frac{5}{6}$			$\frac{1}{100}$		
$\frac{1}{8}$			$1\frac{1}{2}$		
$\frac{3}{8}$					

N

$+ - \times \div$ **Mixed Numbers**

For use with Lesson 64

Name _____

Time _____

Simplify these expressions. Reduce the answers.

$3 + 1\frac{2}{3} =$	$3 - 1\frac{2}{3} =$	$3 \times 1\frac{2}{3} =$	$3 \div 1\frac{2}{3} =$
$1\frac{2}{3} + 1\frac{1}{2} =$	$1\frac{2}{3} - 1\frac{1}{2} =$	$1\frac{2}{3} \times 1\frac{1}{2} =$	$1\frac{2}{3} \div 1\frac{1}{2} =$
$2\frac{1}{2} + 1\frac{2}{3} =$	$2\frac{1}{2} - 1\frac{2}{3} =$	$2\frac{1}{2} \times 1\frac{2}{3} =$	$2\frac{1}{2} \div 1\frac{2}{3} =$
$4\frac{1}{2} + 2\frac{1}{4} =$	$4\frac{1}{2} - 2\frac{1}{4} =$	$4\frac{1}{2} \times 2\frac{1}{4} =$	$4\frac{1}{2} \div 2\frac{1}{4} =$
$6\frac{2}{3} + 3\frac{3}{4} =$	$6\frac{2}{3} - 3\frac{3}{4} =$	$6\frac{2}{3} \times 3\frac{3}{4} =$	$3\frac{3}{4} \div 6\frac{2}{3} =$

M **Metric Conversions**
For use with Lesson 65

Name _____

Time _____

Complete each equivalence.

1. 2 meters = _____ centimeters

2. 1.5 kilometers = _____ meters

3. 2.54 centimeters = _____ millimeters

4. 125 centimeters = _____ meters

5. 75 millimeters = _____ centimeters

6. 0.8 meter = _____ millimeters

7. 10 kilometers = _____ meters

8. 0.1 kilometer = _____ meters

9. 5000 meters = _____ kilometers

10. 50 centimeters = _____ meter

11. 50 centimeters = _____ millimeters

12. 2 liters = _____ milliliters

13. 250 milliliters = _____ liter

14. 4 kilograms = _____ grams

15. 2.5 grams = _____ milligrams

16. 500 milligrams = _____ gram

17. 0.5 kilogram = _____ grams

18. Two liters of water has a volume of

_____ cubic centimeters and a mass

of _____ kilograms.

Record the factor indicated by each prefix.

	Prefix	Factor
19.	kilo-	
20.	hecto-	
21.	deka-	
	(unit)	1
22.	deci-	
23.	centi-	
24.	milli-	

L — Fraction-Decimal-Percent Equivalents
For use with Test 12

Name _____

Time _____

Write each fraction as a decimal and as a percent. Write repeating decimals with a bar over the repetend.

Fraction	Decimal	Percent	Fraction	Decimal	Percent
$\frac{1}{2}$			$\frac{5}{8}$		
$\frac{1}{3}$			$\frac{7}{8}$		
$\frac{2}{3}$			$\frac{1}{9}$		
$\frac{1}{4}$			$\frac{1}{10}$		
$\frac{3}{4}$			$\frac{3}{10}$		
$\frac{1}{5}$			$\frac{7}{10}$		
$\frac{2}{5}$			$\frac{9}{10}$		
$\frac{3}{5}$			$\frac{1}{20}$		
$\frac{4}{5}$			$\frac{1}{25}$		
$\frac{1}{6}$			$\frac{1}{50}$		
$\frac{5}{6}$			$\frac{1}{100}$		
$\frac{1}{8}$			$1\frac{1}{2}$		
$\frac{3}{8}$					

+ − × ÷ Mixed Numbers

For use with Lesson 66

Name _____

Time _____

Simplify these expressions. Reduce the answers.

$3 + 1\frac{2}{3} =$	$3 - 1\frac{2}{3} =$	$3 \times 1\frac{2}{3} =$	$3 \div 1\frac{2}{3} =$
$1\frac{2}{3} + 1\frac{1}{2} =$	$1\frac{2}{3} - 1\frac{1}{2} =$	$1\frac{2}{3} \times 1\frac{1}{2} =$	$1\frac{2}{3} \div 1\frac{1}{2} =$
$2\frac{1}{2} + 1\frac{2}{3} =$	$2\frac{1}{2} - 1\frac{2}{3} =$	$2\frac{1}{2} \times 1\frac{2}{3} =$	$2\frac{1}{2} \div 1\frac{2}{3} =$
$4\frac{1}{2} + 2\frac{1}{4} =$	$4\frac{1}{2} - 2\frac{1}{4} =$	$4\frac{1}{2} \times 2\frac{1}{4} =$	$4\frac{1}{2} \div 2\frac{1}{4} =$
$6\frac{2}{3} + 3\frac{3}{4} =$	$6\frac{2}{3} - 3\frac{3}{4} =$	$6\frac{2}{3} \times 3\frac{3}{4} =$	$3\frac{3}{4} \div 6\frac{2}{3} =$

M

Metric Conversions
For use with Lesson 67

Name _____

Time _____

Complete each equivalence.

1. 2 meters = _____ centimeters

2. 1.5 kilometers = _____ meters

3. 2.54 centimeters = _____ millimeters

4. 125 centimeters = _____ meters

5. 75 millimeters = _____ centimeters

6. 0.8 meter = _____ millimeters

7. 10 kilometers = _____ meters

8. 0.1 kilometer = _____ meters

9. 5000 meters = _____ kilometers

10. 50 centimeters = _____ meter

11. 50 centimeters = _____ millimeters

12. 2 liters = _____ milliliters

13. 250 milliliters = _____ liter

14. 4 kilograms = _____ grams

15. 2.5 grams = _____ milligrams

16. 500 milligrams = _____ gram

17. 0.5 kilogram = _____ grams

18. Two liters of water has a volume of

_____ cubic centimeters and a mass

of _____ kilograms.

Record the factor indicated by each prefix.

	Prefix	Factor
19.	kilo-	
20.	hecto-	
21.	deka-	
	(unit)	1
22.	deci-	
23.	centi-	
24.	milli-	

| N | $+ - \times \div$ **Mixed Numbers**
For use with Lesson 68 | Name _____
Time _____ |

Simplify these expressions. Reduce the answers.

$3 + 1\frac{2}{3} =$	$3 - 1\frac{2}{3} =$	$3 \times 1\frac{2}{3} =$	$3 \div 1\frac{2}{3} =$
$1\frac{2}{3} + 1\frac{1}{2} =$	$1\frac{2}{3} - 1\frac{1}{2} =$	$1\frac{2}{3} \times 1\frac{1}{2} =$	$1\frac{2}{3} \div 1\frac{1}{2} =$
$2\frac{1}{2} + 1\frac{2}{3} =$	$2\frac{1}{2} - 1\frac{2}{3} =$	$2\frac{1}{2} \times 1\frac{2}{3} =$	$2\frac{1}{2} \div 1\frac{2}{3} =$
$4\frac{1}{2} + 2\frac{1}{4} =$	$4\frac{1}{2} - 2\frac{1}{4} =$	$4\frac{1}{2} \times 2\frac{1}{4} =$	$4\frac{1}{2} \div 2\frac{1}{4} =$
$6\frac{2}{3} + 3\frac{3}{4} =$	$6\frac{2}{3} - 3\frac{3}{4} =$	$6\frac{2}{3} \times 3\frac{3}{4} =$	$3\frac{3}{4} \div 6\frac{2}{3} =$

M	**Metric Conversions**	Name _____
	For use with Lesson 69	Time _____

Complete each equivalence.

1. 2 meters = _____ centimeters

2. 1.5 kilometers = _____ meters

3. 2.54 centimeters = _____ millimeters

4. 125 centimeters = _____ meters

5. 75 millimeters = _____ centimeters

6. 0.8 meter = _____ millimeters

7. 10 kilometers = _____ meters

8. 0.1 kilometer = _____ meters

9. 5000 meters = _____ kilometers

10. 50 centimeters = _____ meter

11. 50 centimeters = _____ millimeters

12. 2 liters = _____ milliliters

13. 250 milliliters = _____ liter

14. 4 kilograms = _____ grams

15. 2.5 grams = _____ milligrams

16. 500 milligrams = _____ gram

17. 0.5 kilogram = _____ grams

18. Two liters of water has a volume of

_____ cubic centimeters and a mass

of _____ kilograms.

Record the factor indicated by each prefix.

	Prefix	**Factor**
19.	kilo-	
20.	hecto-	
21.	deka-	
	(unit)	1
22.	deci-	
23.	centi-	
24.	milli-	

N

$+ - \times \div$ Mixed Numbers
For use with Lesson 70

Name _____

Time _____

Simplify these expressions. Reduce the answers.

$3 + 1\frac{2}{3} =$	$3 - 1\frac{2}{3} =$	$3 \times 1\frac{2}{3} =$	$3 \div 1\frac{2}{3} =$
$1\frac{2}{3} + 1\frac{1}{2} =$	$1\frac{2}{3} - 1\frac{1}{2} =$	$1\frac{2}{3} \times 1\frac{1}{2} =$	$1\frac{2}{3} \div 1\frac{1}{2} =$
$2\frac{1}{2} + 1\frac{2}{3} =$	$2\frac{1}{2} - 1\frac{2}{3} =$	$2\frac{1}{2} \times 1\frac{2}{3} =$	$2\frac{1}{2} \div 1\frac{2}{3} =$
$4\frac{1}{2} + 2\frac{1}{4} =$	$4\frac{1}{2} - 2\frac{1}{4} =$	$4\frac{1}{2} \times 2\frac{1}{4} =$	$4\frac{1}{2} \div 2\frac{1}{4} =$
$6\frac{2}{3} + 3\frac{3}{4} =$	$6\frac{2}{3} - 3\frac{3}{4} =$	$6\frac{2}{3} \times 3\frac{3}{4} =$	$3\frac{3}{4} \div 6\frac{2}{3} =$

M

Metric Conversions
For use with Test 13

Name _____

Time _____

Complete each equivalence.

1. 2 meters = _____ centimeters

2. 1.5 kilometers = _____ meters

3. 2.54 centimeters = _____ millimeters

4. 125 centimeters = _____ meters

5. 75 millimeters = _____ centimeters

6. 0.8 meter = _____ millimeters

7. 10 kilometers = _____ meters

8. 0.1 kilometer = _____ meters

9. 5000 meters = _____ kilometers

10. 50 centimeters = _____ meter

11. 50 centimeters = _____ millimeters

12. 2 liters = _____ milliliters

13. 250 milliliters = _____ liter

14. 4 kilograms = _____ grams

15. 2.5 grams = _____ milligrams

16. 500 milligrams = _____ gram

17. 0.5 kilogram = _____ grams

18. Two liters of water has a volume of

_____ cubic centimeters and a mass

of _____ kilograms.

Record the factor indicated by each prefix.

	Prefix	Factor
19.	kilo-	
20.	hecto-	
21.	deka-	
	(unit)	1
22.	deci-	
23.	centi-	
24.	milli-	

O

Classifying Quadrilaterals and Triangles

For use with Lesson 71

Name _____

Time _____

Select from the words at the bottom of the page to describe each figure.

1.

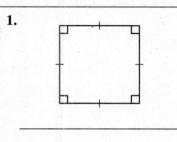

_____

_____

_____

_____

2.

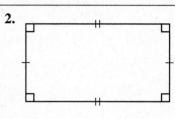

_____

_____

3.

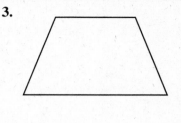

_____

4.

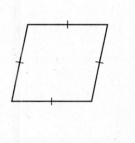

_____

_____

5.

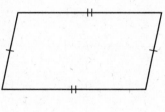

_____

6.

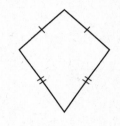

_____

7.

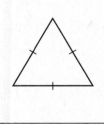

_____

_____

8.

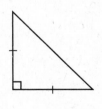

_____

_____

9.

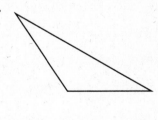

_____

_____

kite	rectangle	isosceles triangle	right triangle
trapezoid	rhombus	scalene triangle	acute triangle
parallelogram	square	equilateral triangle	obtuse triangle

N

+ − × ÷ Mixed Numbers
For use with Lesson 72

Name _____

Time _____

Simplify these expressions. Reduce the answers.

$3 + 1\frac{2}{3} =$	$3 - 1\frac{2}{3} =$	$3 \times 1\frac{2}{3} =$	$3 \div 1\frac{2}{3} =$
$1\frac{2}{3} + 1\frac{1}{2} =$	$1\frac{2}{3} - 1\frac{1}{2} =$	$1\frac{2}{3} \times 1\frac{1}{2} =$	$1\frac{2}{3} \div 1\frac{1}{2} =$
$2\frac{1}{2} + 1\frac{2}{3} =$	$2\frac{1}{2} - 1\frac{2}{3} =$	$2\frac{1}{2} \times 1\frac{2}{3} =$	$2\frac{1}{2} \div 1\frac{2}{3} =$
$4\frac{1}{2} + 2\frac{1}{4} =$	$4\frac{1}{2} - 2\frac{1}{4} =$	$4\frac{1}{2} \times 2\frac{1}{4} =$	$4\frac{1}{2} \div 2\frac{1}{4} =$
$6\frac{2}{3} + 3\frac{3}{4} =$	$6\frac{2}{3} - 3\frac{3}{4} =$	$6\frac{2}{3} \times 3\frac{3}{4} =$	$3\frac{3}{4} \div 6\frac{2}{3} =$

Classifying Quadrilaterals and Triangles

For use with Lesson 73

Name _____

Time _____

Select from the words at the bottom of the page to describe each figure.

1.	**2.**	**3.**
4.	**5.**	**6.**
7.	**8.**	**9.**

kite	rectangle	isosceles triangle	right triangle
trapezoid	rhombus	scalene triangle	acute triangle
parallelogram	square	equilateral triangle	obtuse triangle

N

+ − × ÷ Mixed Numbers

For use with Lesson 74

Name _____

Time _____

Simplify these expressions. Reduce the answers.

$3 + 1\frac{2}{3} =$	$3 - 1\frac{2}{3} =$	$3 \times 1\frac{2}{3} =$	$3 \div 1\frac{2}{3} =$
$1\frac{2}{3} + 1\frac{1}{2} =$	$1\frac{2}{3} - 1\frac{1}{2} =$	$1\frac{2}{3} \times 1\frac{1}{2} =$	$1\frac{2}{3} \div 1\frac{1}{2} =$
$2\frac{1}{2} + 1\frac{2}{3} =$	$2\frac{1}{2} - 1\frac{2}{3} =$	$2\frac{1}{2} \times 1\frac{2}{3} =$	$2\frac{1}{2} \div 1\frac{2}{3} =$
$4\frac{1}{2} + 2\frac{1}{4} =$	$4\frac{1}{2} - 2\frac{1}{4} =$	$4\frac{1}{2} \times 2\frac{1}{4} =$	$4\frac{1}{2} \div 2\frac{1}{4} =$
$6\frac{2}{3} + 3\frac{3}{4} =$	$6\frac{2}{3} - 3\frac{3}{4} =$	$6\frac{2}{3} \times 3\frac{3}{4} =$	$3\frac{3}{4} \div 6\frac{2}{3} =$

Classifying Quadrilaterals and Triangles

For use with Lesson 75

Name _____

Time _____

Select from the words at the bottom of the page to describe each figure.

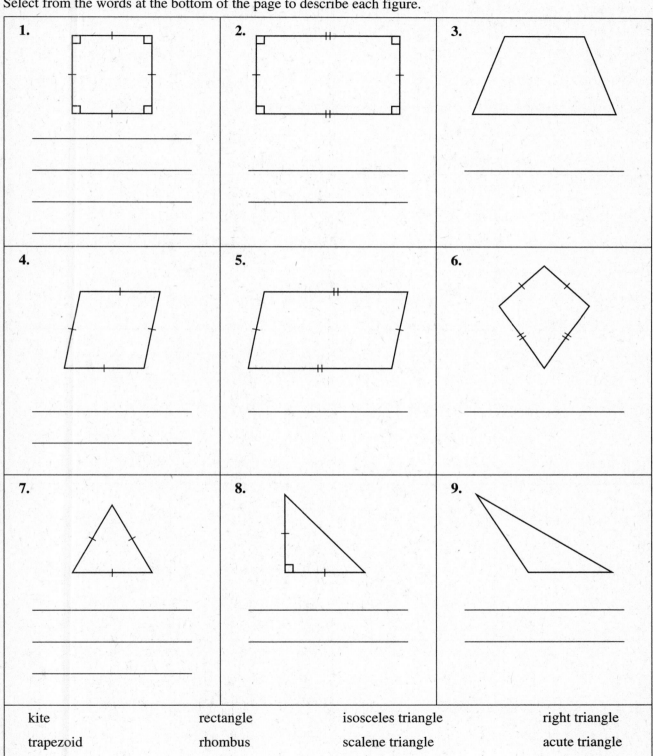

1.

_____
_____
_____
_____

2.

_____
_____

3.

_____

4.

_____
_____

5.

_____

6.

_____

7.

_____
_____
_____

8.

_____
_____

9.

_____
_____

kite	rectangle	isosceles triangle	right triangle
trapezoid	rhombus	scalene triangle	acute triangle
parallelogram	square	equilateral triangle	obtuse triangle

N + − × ÷ **Mixed Numbers**
For use with Test 14

Name _____

Time _____

Simplify these expressions. Reduce the answers.

$3 + 1\frac{2}{3} =$	$3 - 1\frac{2}{3} =$	$3 \times 1\frac{2}{3} =$	$3 \div 1\frac{2}{3} =$
$1\frac{2}{3} + 1\frac{1}{2} =$	$1\frac{2}{3} - 1\frac{1}{2} =$	$1\frac{2}{3} \times 1\frac{1}{2} =$	$1\frac{2}{3} \div 1\frac{1}{2} =$
$2\frac{1}{2} + 1\frac{2}{3} =$	$2\frac{1}{2} - 1\frac{2}{3} =$	$2\frac{1}{2} \times 1\frac{2}{3} =$	$2\frac{1}{2} \div 1\frac{2}{3} =$
$4\frac{1}{2} + 2\frac{1}{4} =$	$4\frac{1}{2} - 2\frac{1}{4} =$	$4\frac{1}{2} \times 2\frac{1}{4} =$	$4\frac{1}{2} \div 2\frac{1}{4} =$
$6\frac{2}{3} + 3\frac{3}{4} =$	$6\frac{2}{3} - 3\frac{3}{4} =$	$6\frac{2}{3} \times 3\frac{3}{4} =$	$3\frac{3}{4} \div 6\frac{2}{3} =$

P

+ − × ÷ Integers
For use with Lesson 76

Name _____

Time _____

Simplify.

$(-8) + (-2) =$	$(-8) - (-2) =$	$(-8)(-2) =$	$\dfrac{-8}{-2} =$
$(-9) + (+3) =$	$(-9) - (+3) =$	$(-9)(+3) =$	$\dfrac{-9}{+3} =$
$12 + (-2) =$	$12 - (-2) =$	$(12)(-2) =$	$\dfrac{12}{-2} =$
$(+12) + (+6) =$	$(+12) - (+6) =$	$(+12)(+6) =$	$\dfrac{+12}{+6} =$
$-20 + (+5) =$	$-20 - (+5) =$	$(-20)(+5) =$	$\dfrac{-20}{+5} =$
$(-15) + (-3) =$	$(-15) - (-3) =$	$(-15)(-3) =$	$\dfrac{-15}{-3} =$
$(+30) + (-6) =$	$(+30) - (-6) =$	$(+30)(-6) =$	$\dfrac{+30}{-6} =$
$(-5) + (-6) + (-2) =$	$(-5) - (-6) - (-2) =$	$(-5)(-6)(-2) =$	$\dfrac{(-5)(-6)}{(-2)} =$

P	**+ − × ÷ Integers**	Name _____
	For use with Lesson 77	Time _____

Simplify.

$(-8) + (-2) =$	$(-8) - (-2) =$	$(-8)(-2) =$	$\dfrac{-8}{-2} =$
$(-9) + (+3) =$	$(-9) - (+3) =$	$(-9)(+3) =$	$\dfrac{-9}{+3} =$
$12 + (-2) =$	$12 - (-2) =$	$(12)(-2) =$	$\dfrac{12}{-2} =$
$(+12) + (+6) =$	$(+12) - (+6) =$	$(+12)(+6) =$	$\dfrac{+12}{+6} =$
$-20 + (+5) =$	$-20 - (+5) =$	$(-20)(+5) =$	$\dfrac{-20}{+5} =$
$(-15) + (-3) =$	$(-15) - (-3) =$	$(-15)(-3) =$	$\dfrac{-15}{-3} =$
$(+30) + (-6) =$	$(+30) - (-6) =$	$(+30)(-6) =$	$\dfrac{+30}{-6} =$
$(-5) + (-6) + (-2) =$	$(-5) - (-6) - (-2) =$	$(-5)(-6)(-2) =$	$\dfrac{(-5)(-6)}{(-2)} =$

Classifying Quadrilaterals and Triangles

For use with Lesson 78

Name _____

Time _____

Select from the words at the bottom of the page to describe each figure.

1.	2.	3.
_____ _____ _____ _____	_____ _____	_____

4.	5.	6.
_____ _____	_____	_____

7.	8.	9.
_____ _____ _____	_____ _____	_____ _____

kite	rectangle	isosceles triangle	right triangle
trapezoid	rhombus	scalene triangle	acute triangle
parallelogram	square	equilateral triangle	obtuse triangle

P

$+ - \times \div$ **Integers**

For use with Lesson 79

Name _____

Time _____

Simplify.

$(-8) + (-2) =$	$(-8) - (-2) =$	$(-8)(-2) =$	$\dfrac{-8}{-2} =$
$(-9) + (+3) =$	$(-9) - (+3) =$	$(-9)(+3) =$	$\dfrac{-9}{+3} =$
$12 + (-2) =$	$12 - (-2) =$	$(12)(-2) =$	$\dfrac{12}{-2} =$
$(+12) + (+6) =$	$(+12) - (+6) =$	$(+12)(+6) =$	$\dfrac{+12}{+6} =$
$-20 + (+5) =$	$-20 - (+5) =$	$(-20)(+5) =$	$\dfrac{-20}{+5} =$
$(-15) + (-3) =$	$(-15) - (-3) =$	$(-15)(-3) =$	$\dfrac{-15}{-3} =$
$(+30) + (-6) =$	$(+30) - (-6) =$	$(+30)(-6) =$	$\dfrac{+30}{-6} =$
$(-5) + (-6) + (-2) =$	$(-5) - (-6) - (-2) =$	$(-5)(-6)(-2) =$	$\dfrac{(-5)(-6)}{(-2)} =$

Classifying Quadrilaterals and Triangles
For use with Lesson 80

Name _____

Time _____

Select from the words at the bottom of the page to describe each figure.

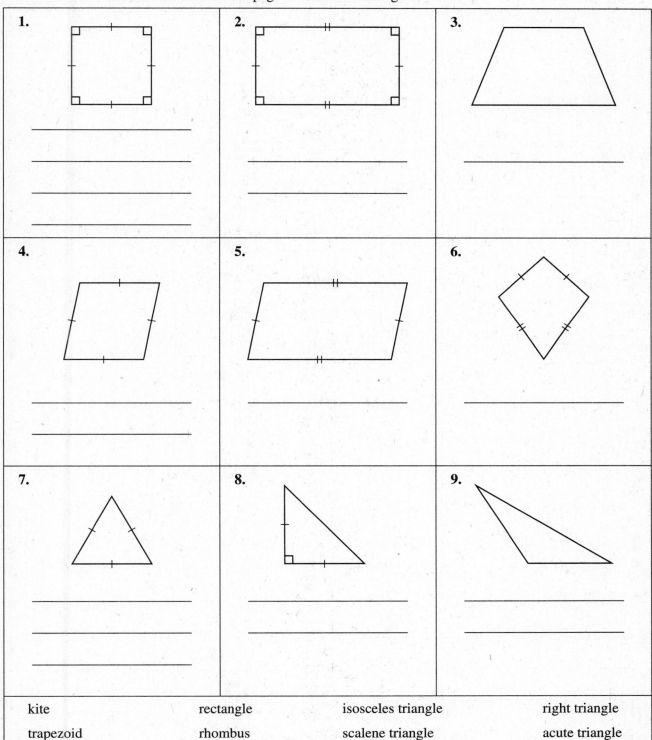

kite	rectangle	isosceles triangle	right triangle
trapezoid	rhombus	scalene triangle	acute triangle
parallelogram	square	equilateral triangle	obtuse triangle

Classifying Quadrilaterals and Triangles

For use with Test 15

Name _____

Time _____

Select from the words at the bottom of the page to describe each figure.

1. 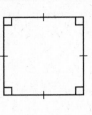 _____ _____	**2.** 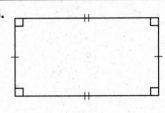 _____ _____	**3.** 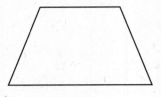 _____
4. 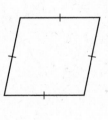 _____ _____	**5.** 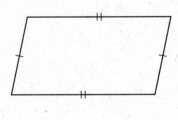 _____	**6.** 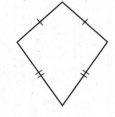 _____
7. _____ _____ _____	**8.** 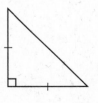 _____ _____	**9.** 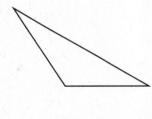 _____ _____

kite	rectangle	isosceles triangle	right triangle
trapezoid	rhombus	scalene triangle	acute triangle
parallelogram	square	equilateral triangle	obtuse triangle

| P | + − × ÷ **Integers** For use with Lesson 81 | Name _____ Time _____ |

Simplify.

(−8) + (−2) =	(−8) − (−2) =	(−8)(−2) =	$\dfrac{-8}{-2} =$
(−9) + (+3) =	(−9) − (+3) =	(−9)(+3) =	$\dfrac{-9}{+3} =$
12 + (−2) =	12 − (−2) =	(12)(−2) =	$\dfrac{12}{-2} =$
(+12) + (+6) =	(+12) − (+6) =	(+12)(+6) =	$\dfrac{+12}{+6} =$
−20 + (+5) =	−20 − (+5) =	(−20)(+5) =	$\dfrac{-20}{+5} =$
(−15) + (−3) =	(−15) − (−3) =	(−15)(−3) =	$\dfrac{-15}{-3} =$
(+30) + (−6) =	(+30) − (−6) =	(+30)(−6) =	$\dfrac{+30}{-6} =$
(−5) + (−6) + (−2) =	(−5) − (−6) − (−2) =	(−5)(−6)(−2) =	$\dfrac{(-5)(-6)}{(-2)} =$

Percent-Decimal-Fraction Equivalents

For use with Lesson 82

Name _____

Time _____

Write each percent as a decimal and as a reduced fraction. Write repeating decimals with a bar over the repetend.

Percent	Decimal	Fraction	Percent	Decimal	Fraction
10%			$62\frac{1}{2}\%$		
90%			20%		
5%			4%		
40%			75%		
$12\frac{1}{2}\%$			$66\frac{2}{3}\%$		
50%			$37\frac{1}{2}\%$		
2%			70%		
30%			1%		
$87\frac{1}{2}\%$			$16\frac{2}{3}\%$		
25%			$83\frac{1}{3}\%$		
80%			$8\frac{1}{3}\%$		
$33\frac{1}{3}\%$			$11\frac{1}{9}\%$		
60%					

R | Area
For use with Lesson 83

Name _____

Time _____

Find the area of each figure. Angles that look like right angles are right angles.

1. 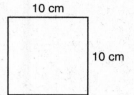 10 cm / 10 cm	**2.** 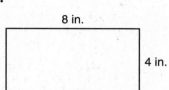 8 in. / 4 in.	**3.** 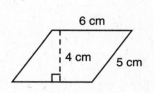 6 cm / 4 cm / 5 cm
4. 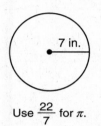 7 in. Use $\frac{22}{7}$ for π.	**5.** 20 cm Use 3.14 for π.	**6.** 10 in. Leave π as π.
7. 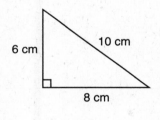 6 cm / 10 cm / 8 cm	**8.** 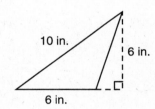 10 in. / 6 in. / 6 in.	**9.** 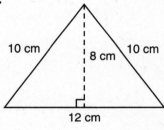 10 cm / 8 cm / 10 cm / 12 cm
10. 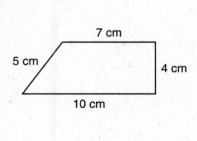 7 cm / 5 cm / 4 cm / 10 cm	**11.** 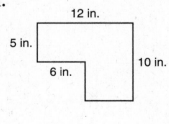 12 in. / 5 in. / 6 in. / 10 in.	**12.** 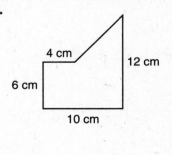 4 cm / 12 cm / 6 cm / 10 cm

R   **Area**
For use with Lesson 84

Name _____

Time _____

Find the area of each figure. Angles that look like right angles are right angles.

1.
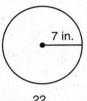
10 cm

10 cm

_____

2.
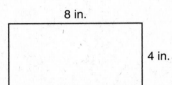
8 in.

4 in.

_____

3.
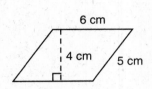
6 cm

4 cm   5 cm

_____

4.
7 in.

Use $\frac{22}{7}$ for π.

_____

5.

20 cm

Use 3.14 for π.

_____

6.

10 in.

Leave π as π.

_____

7.
6 cm   10 cm

8 cm

_____

8.
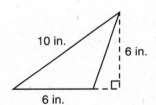
10 in.   6 in.

6 in.

_____

9.
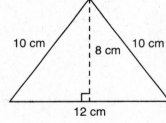
10 cm   8 cm   10 cm

12 cm

_____

10.
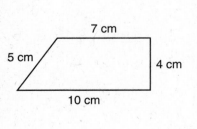
7 cm

5 cm   4 cm

10 cm

_____

11.
12 in.

5 in.   10 in.

6 in.

_____

12.
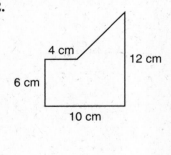
4 cm   12 cm

6 cm

10 cm

_____

P

+ − × ÷ Integers
For use with Lesson 85

Name _____

Time _____

Simplify.

$(-8) + (-2) =$	$(-8) - (-2) =$	$(-8)(-2) =$	$\dfrac{-8}{-2} =$
$(-9) + (+3) =$	$(-9) - (+3) =$	$(-9)(+3) =$	$\dfrac{-9}{+3} =$
$12 + (-2) =$	$12 - (-2) =$	$(12)(-2) =$	$\dfrac{12}{-2} =$
$(+12) + (+6) =$	$(+12) - (+6) =$	$(+12)(+6) =$	$\dfrac{+12}{+6} =$
$-20 + (+5) =$	$-20 - (+5) =$	$(-20)(+5) =$	$\dfrac{-20}{+5} =$
$(-15) + (-3) =$	$(-15) - (-3) =$	$(-15)(-3) =$	$\dfrac{-15}{-3} =$
$(+30) + (-6) =$	$(+30) - (-6) =$	$(+30)(-6) =$	$\dfrac{+30}{-6} =$
$(-5) + (-6) + (-2) =$	$(-5) - (-6) - (-2) =$	$(-5)(-6)(-2) =$	$\dfrac{(-5)(-6)}{(-2)} =$

P   **+ − × ÷ Integers**
For use with Test 16

Name _____

Time _____

Simplify.

$(-8) + (-2) =$	$(-8) - (-2) =$	$(-8)(-2) =$	$\dfrac{-8}{-2} =$
$(-9) + (+3) =$	$(-9) - (+3) =$	$(-9)(+3) =$	$\dfrac{-9}{+3} =$
$12 + (-2) =$	$12 - (-2) =$	$(12)(-2) =$	$\dfrac{12}{-2} =$
$(+12) + (+6) =$	$(+12) - (+6) =$	$(+12)(+6) =$	$\dfrac{+12}{+6} =$
$-20 + (+5) =$	$-20 - (+5) =$	$(-20)(+5) =$	$\dfrac{-20}{+5} =$
$(-15) + (-3) =$	$(-15) - (-3) =$	$(-15)(-3) =$	$\dfrac{-15}{-3} =$
$(+30) + (-6) =$	$(+30) - (-6) =$	$(+30)(-6) =$	$\dfrac{+30}{-6} =$
$(-5) + (-6) + (-2) =$	$(-5) - (-6) - (-2) =$	$(-5)(-6)(-2) =$	$\dfrac{(-5)(-6)}{(-2)} =$

| S | **Scientific Notation**
For use with Lesson 86 |
|---|---|

Name _____

Time _____

Write each number in scientific notation.

186,000 =	0.0005 =
30,500,000 =	36×10^4 =
0.35×10^5 =	48×10^{-3} =
2.5 billion =	15 thousandths =
12 million =	$\dfrac{1}{1,000,000}$ =

Write each number in standard form.

1×10^6 =	1×10^{-6} =
2.4×10^4 =	5×10^{-4} =
4.75×10^5 =	2.5×10^{-3} =
3.125×10^3 =	1.25×10^{-2} =
3.025×10^2 =	1.05×10^{-1} =

R   Area
For use with Lesson 87

Name _____

Time _____

Find the area of each figure. Angles that look like right angles are right angles.

1. 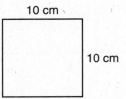 10 cm / 10 cm	**2.** 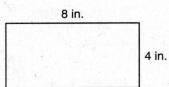 8 in. / 4 in.	**3.** 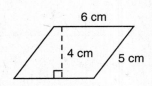 6 cm / 4 cm / 5 cm
4. 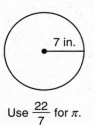 7 in. Use $\frac{22}{7}$ for π.	**5.** 20 cm Use 3.14 for π.	**6.** 10 in. Leave π as π.
7. 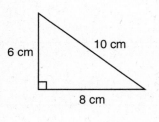 6 cm / 10 cm / 8 cm	**8.** 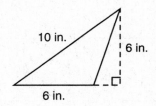 10 in. / 6 in. / 6 in.	**9.** 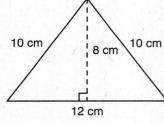 10 cm / 8 cm / 10 cm / 12 cm
10. 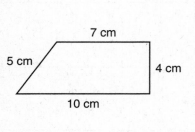 7 cm / 5 cm / 4 cm / 10 cm	**11.** 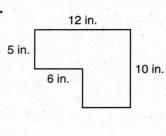 12 in. / 5 in. / 6 in. / 10 in.	**12.** 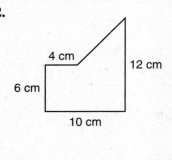 4 cm / 12 cm / 6 cm / 10 cm

S	**Scientific Notation**
	For use with Lesson 88

Name _____

Time _____

Write each number in scientific notation.

186,000 =	0.0005 =
30,500,000 =	36×10^4 =
0.35×10^5 =	48×10^{-3} =
2.5 billion =	15 thousandths =
12 million =	$\dfrac{1}{1,000,000}$ =

Write each number in standard form.

1×10^6 =	1×10^{-6} =
2.4×10^4 =	5×10^{-4} =
4.75×10^5 =	2.5×10^{-3} =
3.125×10^3 =	1.25×10^{-2} =
3.025×10^2 =	1.05×10^{-1} =

R **Area**
For use with Lesson 89

Name _____

Time _____

Find the area of each figure. Angles that look like right angles are right angles.

1.

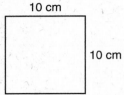

10 cm

10 cm

_____

2.

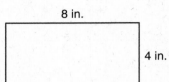

8 in.

4 in.

_____

3.

6 cm

4 cm

5 cm

_____

4.

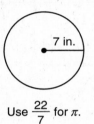

7 in.

Use $\frac{22}{7}$ for π.

_____

5.

20 cm

Use 3.14 for π.

_____

6.

10 in.

Leave π as π.

_____

7.

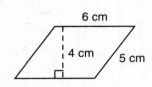

6 cm

10 cm

8 cm

_____

8.

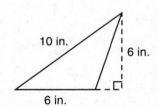

10 in.

6 in.

6 in.

_____

9.

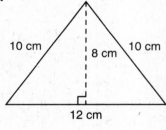

10 cm

8 cm

10 cm

12 cm

_____

10.

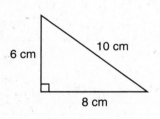

7 cm

5 cm

4 cm

10 cm

_____

11.

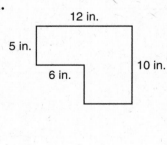

12 in.

5 in.

6 in.

10 in.

_____

12.

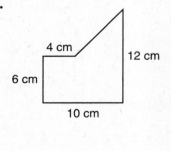

4 cm

12 cm

6 cm

10 cm

_____

130

Percent-Decimal-Fraction Equivalents
For use with Lesson 90

Name _____

Time _____

Write each percent as a decimal and as a reduced fraction. Write repeating decimals with a bar over the repetend.

Percent	Decimal	Fraction
10%		
90%		
5%		
40%		
$12\frac{1}{2}\%$		
50%		
2%		
30%		
$87\frac{1}{2}\%$		
25%		
80%		
$33\frac{1}{3}\%$		
60%		

Percent	Decimal	Fraction
$62\frac{1}{2}\%$		
20%		
4%		
75%		
$66\frac{2}{3}\%$		
$37\frac{1}{2}\%$		
70%		
1%		
$16\frac{2}{3}\%$		
$83\frac{1}{3}\%$		
$8\frac{1}{3}\%$		
$11\frac{1}{9}\%$		

Percent-Decimal-Fraction Equivalents

For use with Test 17

Name _____

Time _____

Write each percent as a decimal and as a reduced fraction. Write repeating decimals with a bar over the repetend.

Percent	Decimal	Fraction
10%		
90%		
5%		
40%		
$12\frac{1}{2}\%$		
50%		
2%		
30%		
$87\frac{1}{2}\%$		
25%		
80%		
$33\frac{1}{3}\%$		
60%		

Percent	Decimal	Fraction
$62\frac{1}{2}\%$		
20%		
4%		
75%		
$66\frac{2}{3}\%$		
$37\frac{1}{2}\%$		
70%		
1%		
$16\frac{2}{3}\%$		
$83\frac{1}{3}\%$		
$8\frac{1}{3}\%$		
$11\frac{1}{9}\%$		

S | Scientific Notation
For use with Lesson 91

Name _____

Time _____

Write each number in scientific notation.

186,000 =	0.0005 =
30,500,000 =	36×10^4 =
0.35×10^5 =	48×10^{-3} =
2.5 billion =	15 thousandths =
12 million =	$\dfrac{1}{1,000,000}$ =

Write each number in standard form.

1×10^6 =	1×10^{-6} =
2.4×10^4 =	5×10^{-4} =
4.75×10^5 =	2.5×10^{-3} =
3.125×10^3 =	1.25×10^{-2} =
3.025×10^2 =	1.05×10^{-1} =

T	**Order of Operations**	Name _____
	For use with Lesson 92	Time _____

Simplify.

$6 + 6 \times 6 - 6 \div 6 =$	$5 + 5^2 + 5 \div 5 - 5 \times 5 =$
$3^2 + \sqrt{4} + 5(6) - 7 + 8 =$	$6 \times 4 \div 2 - 6 \div 2 \times 4 =$
$4 + 2(3 + 5) - 6 \div 2 =$	$8 + 7 \times 6 - (5 + 4) \div 3 + 2 =$
$2 + 2[3 + 4(7 - 5)] =$	$3[10 + (6 - 4) - 3(2 + 1)] =$
$\dfrac{(4)(3)(2)}{4 - 3 + 2} =$	$\sqrt{1^3 + 2^3 + 3^3} =$
$\dfrac{6 + 8(7 - 5) - 2}{4(3) - (4 + 3)} =$	$(2 + 3)^2 + 5[4^2 - 2(3)] =$
$(-3) + (-3)(-3) - (-3) =$	$\sqrt{-3 - (3)(-3) - (-3)} =$
$\dfrac{3(-3) - (-3)(-3)}{(-3) - 3(-3)} =$	$\dfrac{(-3) - (-3) - \sqrt{3(3)}}{3^2 - 3(3) - 3} =$

T | Order of Operations
For use with Lesson 93

Name _____

Time _____

Simplify.

$6 + 6 \times 6 - 6 \div 6 =$	$5 + 5^2 + 5 \div 5 - 5 \times 5 =$
$3^2 + \sqrt{4} + 5(6) - 7 + 8 =$	$6 \times 4 \div 2 - 6 \div 2 \times 4 =$
$4 + 2(3 + 5) - 6 \div 2 =$	$8 + 7 \times 6 - (5 + 4) \div 3 + 2 =$
$2 + 2[3 + 4(7 - 5)] =$	$3[10 + (6 - 4) - 3(2 + 1)] =$
$\dfrac{(4)(3)(2)}{4 - 3 + 2} =$	$\sqrt{1^3 + 2^3 + 3^3} =$
$\dfrac{6 + 8(7 - 5) - 2}{4(3) - (4 + 3)} =$	$(2 + 3)^2 + 5[4^2 - 2(3)] =$
$(-3) + (-3)(-3) - (-3) =$	$\sqrt{-3 - (3)(-3) - (-3)} =$
$\dfrac{3(-3) - (-3)(-3)}{(-3) - 3(-3)} =$	$\dfrac{(-3) - (-3) - \sqrt{3(3)}}{3^2 - 3(3) - 3} =$

S

Scientific Notation
For use with Lesson 94

Name _____

Time _____

Write each number in scientific notation.

186,000 =	0.0005 =
30,500,000 =	36×10^4 =
0.35×10^5 =	48×10^{-3} =
2.5 billion =	15 thousandths =
12 million =	$\dfrac{1}{1,000,000}$ =

Write each number in standard form.

1×10^6 =	1×10^{-6} =
2.4×10^4 =	5×10^{-4} =
4.75×10^5 =	2.5×10^{-3} =
3.125×10^3 =	1.25×10^{-2} =
3.025×10^2 =	1.05×10^{-1} =

<table>
<tr><td>

T

Order of Operations
For use with Lesson 95

</td><td>

Name _____

Time _____

</td></tr>
</table>

Simplify.

$6 + 6 \times 6 - 6 \div 6 =$	$5 + 5^2 + 5 \div 5 - 5 \times 5 =$
$3^2 + \sqrt{4} + 5(6) - 7 + 8 =$	$6 \times 4 \div 2 - 6 \div 2 \times 4 =$
$4 + 2(3 + 5) - 6 \div 2 =$	$8 + 7 \times 6 - (5 + 4) \div 3 + 2 =$
$2 + 2[3 + 4(7 - 5)] =$	$3[10 + (6 - 4) - 3(2 + 1)] =$
$\dfrac{(4)(3)(2)}{4 - 3 + 2} =$	$\sqrt{1^3 + 2^3 + 3^3} =$
$\dfrac{6 + 8(7 - 5) - 2}{4(3) - (4 + 3)} =$	$(2 + 3)^2 + 5[4^2 - 2(3)] =$
$(-3) + (-3)(-3) - (-3) =$	$\sqrt{-3 - (3)(-3) - (-3)} =$
$\dfrac{3(-3) - (-3)(-3)}{(-3) - 3(-3)} =$	$\dfrac{(-3) - (-3) - \sqrt{3(3)}}{3^2 - 3(3) - 3} =$

R **Area**
For use with Test 18

Name _____

Time _____

Find the area of each figure. Angles that look like right angles are right angles.

1.

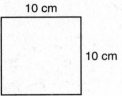

10 cm

10 cm

_____

2.

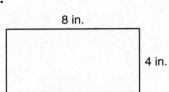

8 in.

4 in.

_____

3.

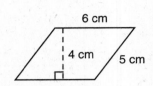

6 cm

4 cm

5 cm

_____

4.

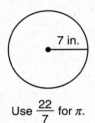

7 in.

Use $\frac{22}{7}$ for π.

_____

5.

20 cm

Use 3.14 for π.

_____

6.

10 in.

Leave π as π.

_____

7.

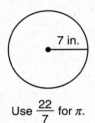

6 cm

10 cm

8 cm

_____

8.

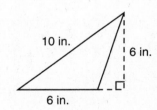

10 in.

6 in.

6 in.

_____

9.

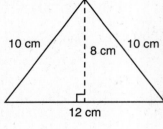

10 cm   8 cm   10 cm

12 cm

_____

10.

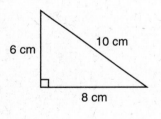

7 cm

5 cm

4 cm

10 cm

_____

11.

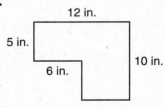

12 in.

5 in.

6 in.

10 in.

_____

12.

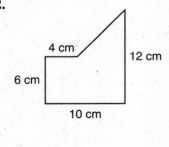

4 cm

12 cm

6 cm

10 cm

_____

Two-Step Equations
For use with Lesson 96

Name _____

Time _____

Complete each step to solve these equations.

$2x + 5 = 45$ $2x =$ $x =$	$3y + 4 = 22$ $3y =$ $y =$	$6w + 8 = 50$ $6w =$ $w =$
$5n - 3 = 32$ $5n =$ $n =$	$3m - 7 = 26$ $3m =$ $m =$	$8p - 9 = 47$ $8p =$ $p =$
$15 = 3a - 6$ $= 3a$ $= a$	$24 = 3b + 6$ $= 3b$ $= b$	$45 = 5c - 10$ $= 5c$ $= c$
$-2x + 9 = 25$ $-2x =$ $x =$	$\frac{3}{4}m + 12 = 36$ $\frac{3}{4}m =$ $m =$	$0.5w - 1.5 = 4.5$ $0.5w =$ $w =$
$-\frac{2}{3}n - 6 = 18$ $-\frac{2}{3}n =$ $n =$	$25 = 10 - 5y$ $= -5y$ $= y$	$-0.3f + 1.2 = 4.8$ $-0.3f =$ $f =$

T

Order of Operations
For use with Lesson 97

Name _____

Time _____

Simplify.

$6 + 6 \times 6 - 6 \div 6 =$	$5 + 5^2 + 5 \div 5 - 5 \times 5 =$
$3^2 + \sqrt{4} + 5(6) - 7 + 8 =$	$6 \times 4 \div 2 - 6 \div 2 \times 4 =$
$4 + 2(3 + 5) - 6 \div 2 =$	$8 + 7 \times 6 - (5 + 4) \div 3 + 2 =$
$2 + 2[3 + 4(7 - 5)] =$	$3[10 + (6 - 4) - 3(2 + 1)] =$
$\dfrac{(4)(3)(2)}{4 - 3 + 2} =$	$\sqrt{1^3 + 2^3 + 3^3} =$
$\dfrac{6 + 8(7 - 5) - 2}{4(3) - (4 + 3)} =$	$(2 + 3)^2 + 5[4^2 - 2(3)] =$
$(-3) + (-3)(-3) - (-3) =$	$\sqrt{-3 - (3)(-3) - (-3)} =$
$\dfrac{3(-3) - (-3)(-3)}{(-3) - 3(-3)} =$	$\dfrac{(-3) - (-3) - \sqrt{3(3)}}{3^2 - 3(3) - 3} =$

Saxon Math 8/7—Homeschool

Two-Step Equations

For use with Lesson 98

Name _____

Time _____

Complete each step to solve these equations.

$2x + 5 = 45$ $2x =$ $x =$	$3y + 4 = 22$ $3y =$ $y =$	$6w + 8 = 50$ $6w =$ $w =$
$5n - 3 = 32$ $5n =$ $n =$	$3m - 7 = 26$ $3m =$ $m =$	$8p - 9 = 47$ $8p =$ $p =$
$15 = 3a - 6$ $= 3a$ $= a$	$24 = 3b + 6$ $= 3b$ $= b$	$45 = 5c - 10$ $= 5c$ $= c$
$-2x + 9 = 25$ $-2x =$ $x =$	$\frac{3}{4}m + 12 = 36$ $\frac{3}{4}m =$ $m =$	$0.5w - 1.5 = 4.5$ $0.5w =$ $w =$
$-\frac{2}{3}n - 6 = 18$ $-\frac{2}{3}n =$ $n =$	$25 = 10 - 5y$ $= -5y$ $= y$	$-0.3f + 1.2 = 4.8$ $-0.3f =$ $f =$

T	**Order of Operations** *For use with Lesson 99*

Name _____

Time _____

Simplify.

$6 + 6 \times 6 - 6 \div 6 =$	$5 + 5^2 + 5 \div 5 - 5 \times 5 =$
$3^2 + \sqrt{4} + 5(6) - 7 + 8 =$	$6 \times 4 \div 2 - 6 \div 2 \times 4 =$
$4 + 2(3 + 5) - 6 \div 2 =$	$8 + 7 \times 6 - (5 + 4) \div 3 + 2 =$
$2 + 2[3 + 4(7 - 5)] =$	$3[10 + (6 - 4) - 3(2 + 1)] =$
$\dfrac{(4)(3)(2)}{4 - 3 + 2} =$	$\sqrt{1^3 + 2^3 + 3^3} =$
$\dfrac{6 + 8(7 - 5) - 2}{4(3) - (4 + 3)} =$	$(2 + 3)^2 + 5[4^2 - 2(3)] =$
$(-3) + (-3)(-3) - (-3) =$	$\sqrt{-3 - (3)(-3) - (-3)} =$
$\dfrac{3(-3) - (-3)(-3)}{(-3) - 3(-3)} =$	$\dfrac{(-3) - (-3) - \sqrt{3(3)}}{3^2 - 3(3) - 3} =$

Two-Step Equations
For use with Lesson 100

Name _____

Time _____

Complete each step to solve these equations.

$2x + 5 = 45$ $2x =$ $x =$	$3y + 4 = 22$ $3y =$ $y =$	$6w + 8 = 50$ $6w =$ $w =$
$5n - 3 = 32$ $5n =$ $n =$	$3m - 7 = 26$ $3m =$ $m =$	$8p - 9 = 47$ $8p =$ $p =$
$15 = 3a - 6$ $= 3a$ $= a$	$24 = 3b + 6$ $= 3b$ $= b$	$45 = 5c - 10$ $= 5c$ $= c$
$-2x + 9 = 25$ $-2x =$ $x =$	$\frac{3}{4}m + 12 = 36$ $\frac{3}{4}m =$ $m =$	$0.5w - 1.5 = 4.5$ $0.5w =$ $w =$
$-\frac{2}{3}n - 6 = 18$ $-\frac{2}{3}n =$ $n =$	$25 = 10 - 5y$ $= -5y$ $= y$	$-0.3f + 1.2 = 4.8$ $-0.3f =$ $f =$

FACTS PRACTICE TEST

S	**Scientific Notation**	Name _____
	For use with Test 19	Time _____

Write each number in scientific notation.

186,000 =	0.0005 =
30,500,000 =	36×10^4 =
0.35×10^5 =	48×10^{-3} =
2.5 billion =	15 thousandths =
12 million =	$\dfrac{1}{1,000,000}$ =

Write each number in standard form.

1×10^6 =	1×10^{-6} =
2.4×10^4 =	5×10^{-4} =
4.75×10^5 =	2.5×10^{-3} =
3.125×10^3 =	1.25×10^{-2} =
3.025×10^2 =	1.05×10^{-1} =

7 Probability Experiment

For use with Investigation 10

Name _____

Section A: Possible outcomes of rolling a pair of dice

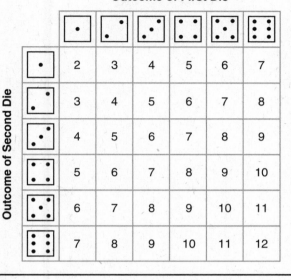

Section B: Theoretical outcomes of 36 rolls of a pair of dice

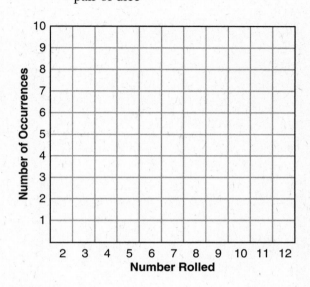

Section C: Actual results of rolling a pair of dice 36 times

Number Rolled	Tally
2	
3	
4	
5	
6	
7	
8	
9	
10	
11	
12	

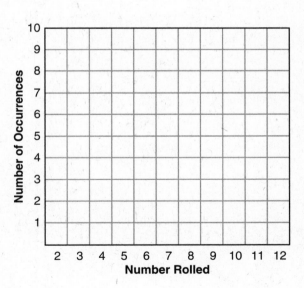

Section D: Possible reasons for a difference between theoretical outcome and actual results

Percent-Decimal-Fraction Equivalents

For use with Lesson 101

Name _____

Time _____

Write each percent as a decimal and as a reduced fraction. Write repeating decimals with a bar over the repetend.

Percent	Decimal	Fraction
10%		
90%		
5%		
40%		
$12\frac{1}{2}\%$		
50%		
2%		
30%		
$87\frac{1}{2}\%$		
25%		
80%		
$33\frac{1}{3}\%$		
60%		

Percent	Decimal	Fraction
$62\frac{1}{2}\%$		
20%		
4%		
75%		
$66\frac{2}{3}\%$		
$37\frac{1}{2}\%$		
70%		
1%		
$16\frac{2}{3}\%$		
$83\frac{1}{3}\%$		
$8\frac{1}{3}\%$		
$11\frac{1}{9}\%$		

U

Two-Step Equations
For use with Lesson 102

Name _____

Time _____

Complete each step to solve these equations.

$2x + 5 = 45$ $2x =$ $x =$	$3y + 4 = 22$ $3y =$ $y =$	$6w + 8 = 50$ $6w =$ $w =$
$5n - 3 = 32$ $5n =$ $n =$	$3m - 7 = 26$ $3m =$ $m =$	$8p - 9 = 47$ $8p =$ $p =$
$15 = 3a - 6$ $= 3a$ $= a$	$24 = 3b + 6$ $= 3b$ $= b$	$45 = 5c - 10$ $= 5c$ $= c$
$-2x + 9 = 25$ $-2x =$ $x =$	$\frac{3}{4}m + 12 = 36$ $\frac{3}{4}m =$ $m =$	$0.5w - 1.5 = 4.5$ $0.5w =$ $w =$
$-\frac{2}{3}n - 6 = 18$ $-\frac{2}{3}n =$ $n =$	$25 = 10 - 5y$ $= -5y$ $= y$	$-0.3f + 1.2 = 4.8$ $-0.3f =$ $f =$

Saxon Math 8/7—Homeschool

Two-Step Equations
For use with Lesson 103

Name _____

Time _____

Complete each step to solve these equations.

$2x + 5 = 45$ $2x =$ $x =$	$3y + 4 = 22$ $3y =$ $y =$	$6w + 8 = 50$ $6w =$ $w =$
$5n - 3 = 32$ $5n =$ $n =$	$3m - 7 = 26$ $3m =$ $m =$	$8p - 9 = 47$ $8p =$ $p =$
$15 = 3a - 6$ $= 3a$ $= a$	$24 = 3b + 6$ $= 3b$ $= b$	$45 = 5c - 10$ $= 5c$ $= c$
$-2x + 9 = 25$ $-2x =$ $x =$	$\frac{3}{4}m + 12 = 36$ $\frac{3}{4}m =$ $m =$	$0.5w - 1.5 = 4.5$ $0.5w =$ $w =$
$-\frac{2}{3}n - 6 = 18$ $-\frac{2}{3}n =$ $n =$	$25 = 10 - 5y$ $= -5y$ $= y$	$-0.3f + 1.2 = 4.8$ $-0.3f =$ $f =$

V

+ − × ÷ Algebraic Terms
For use with Lesson 104

Name _____

Time _____

Simplify.

$6x + 2x =$	$6x - 2x =$	$(6x)(2x) =$	$\dfrac{6x}{2x} =$
$6xy + 2xy =$	$6xy - 2xy =$	$6xy(2xy) =$	$\dfrac{6xy}{2xy} =$
$x + y + x =$	$x + y - x =$	$(x)(y)(-x) =$	$\dfrac{xy}{x} =$
$3x + x + 3 =$	$3x - x - 3 =$	$(3x)(-x)(-3) =$	$\dfrac{(2x)(8xy)}{4y} =$
$3x + 2y + x - y =$		$5xy - 2x + xy - x =$	

Percent-Decimal-Fraction Equivalents
For use with Lesson 105

Name _____

Time _____

Write each percent as a decimal and as a reduced fraction. Write repeating decimals with a bar over the repetend.

Percent	Decimal	Fraction
10%		
90%		
5%		
40%		
$12\frac{1}{2}\%$		
50%		
2%		
30%		
$87\frac{1}{2}\%$		
25%		
80%		
$33\frac{1}{3}\%$		
60%		

Percent	Decimal	Fraction
$62\frac{1}{2}\%$		
20%		
4%		
75%		
$66\frac{2}{3}\%$		
$37\frac{1}{2}\%$		
70%		
1%		
$16\frac{2}{3}\%$		
$83\frac{1}{3}\%$		
$8\frac{1}{3}\%$		
$11\frac{1}{9}\%$		

| T | **Order of Operations**
For use with Test 20 |
|---|---|

Name _____

Time _____

Simplify.

$6 + 6 \times 6 - 6 \div 6 =$	$5 + 5^2 + 5 \div 5 - 5 \times 5 =$
$3^2 + \sqrt{4} + 5(6) - 7 + 8 =$	$6 \times 4 \div 2 - 6 \div 2 \times 4 =$
$4 + 2(3 + 5) - 6 \div 2 =$	$8 + 7 \times 6 - (5 + 4) \div 3 + 2 =$
$2 + 2[3 + 4(7 - 5)] =$	$3[10 + (6 - 4) - 3(2 + 1)] =$
$\dfrac{(4)(3)(2)}{4 - 3 + 2} =$	$\sqrt{1^3 + 2^3 + 3^3} =$
$\dfrac{6 + 8(7 - 5) - 2}{4(3) - (4 + 3)} =$	$(2 + 3)^2 + 5[4^2 - 2(3)] =$
$(-3) + (-3)(-3) - (-3) =$	$\sqrt{-3 - (3)(-3) - (-3)} =$
$\dfrac{3(-3) - (-3)(-3)}{(-3) - 3(-3)} =$	$\dfrac{(-3) - (-3) - \sqrt{3(3)}}{3^2 - 3(3) - 3} =$

V

+ − × ÷ Algebraic Terms
For use with Lesson 106

Name _____

Time _____

Simplify.

$6x + 2x =$	$6x - 2x =$	$(6x)(2x) =$	$\dfrac{6x}{2x} =$
$6xy + 2xy =$	$6xy - 2xy =$	$6xy(2xy) =$	$\dfrac{6xy}{2xy} =$
$x + y + x =$	$x + y - x =$	$(x)(y)(-x) =$	$\dfrac{xy}{x} =$
$3x + x + 3 =$	$3x - x - 3 =$	$(3x)(-x)(-3) =$	$\dfrac{(2x)(8xy)}{4y} =$
$3x + 2y + x - y =$		$5xy - 2x + xy - x =$	

Q Percent-Decimal-Fraction Equivalents
For use with Lesson 107

Name _____

Time _____

Write each percent as a decimal and as a reduced fraction. Write repeating decimals with a bar over the repetend.

Percent	Decimal	Fraction
10%		
90%		
5%		
40%		
$12\frac{1}{2}\%$		
50%		
2%		
30%		
$87\frac{1}{2}\%$		
25%		
80%		
$33\frac{1}{3}\%$		
60%		

Percent	Decimal	Fraction
$62\frac{1}{2}\%$		
20%		
4%		
75%		
$66\frac{2}{3}\%$		
$37\frac{1}{2}\%$		
70%		
1%		
$16\frac{2}{3}\%$		
$83\frac{1}{3}\%$		
$8\frac{1}{3}\%$		
$11\frac{1}{9}\%$		

8 | Slope

For use with Lesson 107

Name _____

Calculate the slope of each line **a–h** below.

1. Slope of line **a:** _____

2. Slope of line **b:** _____

3. Slope of line **c:** _____

4. Slope of line **d:** _____

5. Slope of line **e:** _____

6. Slope of line **f:** _____

7. Slope of line **g:** _____

8. Slope of line **h:** _____

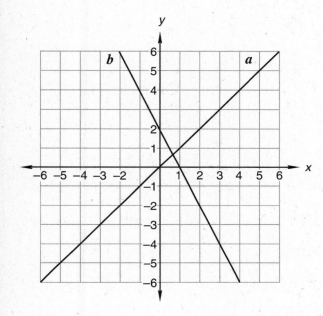

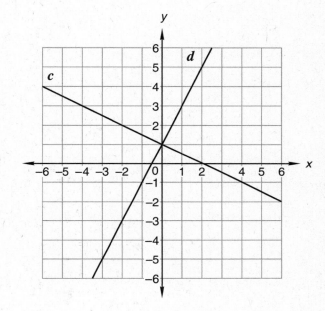

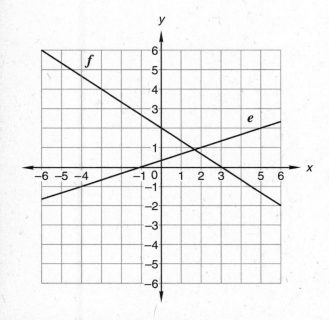

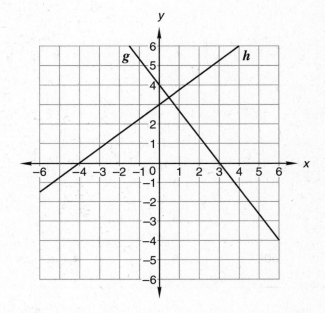

V

$+ - \times \div$ **Algebraic Terms**
For use with Lesson 108

Name _____

Time _____

Simplify.

$6x + 2x =$	$6x - 2x =$	$(6x)(2x) =$	$\dfrac{6x}{2x} =$
$6xy + 2xy =$	$6xy - 2xy =$	$6xy(2xy) =$	$\dfrac{6xy}{2xy} =$
$x + y + x =$	$x + y - x =$	$(x)(y)(-x) =$	$\dfrac{xy}{x} =$
$3x + x + 3 =$	$3x - x - 3 =$	$(3x)(-x)(-3) =$	$\dfrac{(2x)(8xy)}{4y} =$
$3x + 2y + x - y =$		$5xy - 2x + xy - x =$	

$+ - \times \div$ **Algebraic Terms**
For use with Lesson 109

Name _____

Time _____

Simplify.

$6x + 2x =$	$6x - 2x =$	$(6x)(2x) =$	$\dfrac{6x}{2x} =$
$6xy + 2xy =$	$6xy - 2xy =$	$6xy(2xy) =$	$\dfrac{6xy}{2xy} =$
$x + y + x =$	$x + y - x =$	$(x)(y)(-x) =$	$\dfrac{xy}{x} =$
$3x + x + 3 =$	$3x - x - 3 =$	$(3x)(-x)(-3) =$	$\dfrac{(2x)(8xy)}{4y} =$
$3x + 2y + x - y =$		$5xy - 2x + xy - x =$	

Percent-Decimal-Fraction Equivalents
For use with Lesson 110

Name _____

Time _____

Write each percent as a decimal and as a reduced fraction. Write repeating decimals with a bar over the repetend.

Percent	Decimal	Fraction	Percent	Decimal	Fraction
10%			$62\frac{1}{2}\%$		
90%			20%		
5%			4%		
40%			75%		
$12\frac{1}{2}\%$			$66\frac{2}{3}\%$		
50%			$37\frac{1}{2}\%$		
2%			70%		
30%			1%		
$87\frac{1}{2}\%$			$16\frac{2}{3}\%$		
25%			$83\frac{1}{3}\%$		
80%			$8\frac{1}{3}\%$		
$33\frac{1}{3}\%$			$11\frac{1}{9}\%$		
60%					

U Two-Step Equations
For use with Test 21

Name _____

Time _____

Complete each step to solve these equations.

$2x + 5 = 45$ $2x =$ $x =$	$3y + 4 = 22$ $3y =$ $y =$	$6w + 8 = 50$ $6w =$ $w =$
$5n - 3 = 32$ $5n =$ $n =$	$3m - 7 = 26$ $3m =$ $m =$	$8p - 9 = 47$ $8p =$ $p =$
$15 = 3a - 6$ $= 3a$ $= a$	$24 = 3b + 6$ $= 3b$ $= b$	$45 = 5c - 10$ $= 5c$ $= c$
$-2x + 9 = 25$ $-2x =$ $x =$	$\frac{3}{4}m + 12 = 36$ $\frac{3}{4}m =$ $m =$	$0.5w - 1.5 = 4.5$ $0.5w =$ $w =$
$-\frac{2}{3}n - 6 = 18$ $-\frac{2}{3}n =$ $n =$	$25 = 10 - 5y$ $= -5y$ $= y$	$-0.3f + 1.2 = 4.8$ $-0.3f =$ $f =$

9 Square Centimeter Grid

For use with Investigation 11

10 | Square Centimeter Grid
For use with Investigation 11

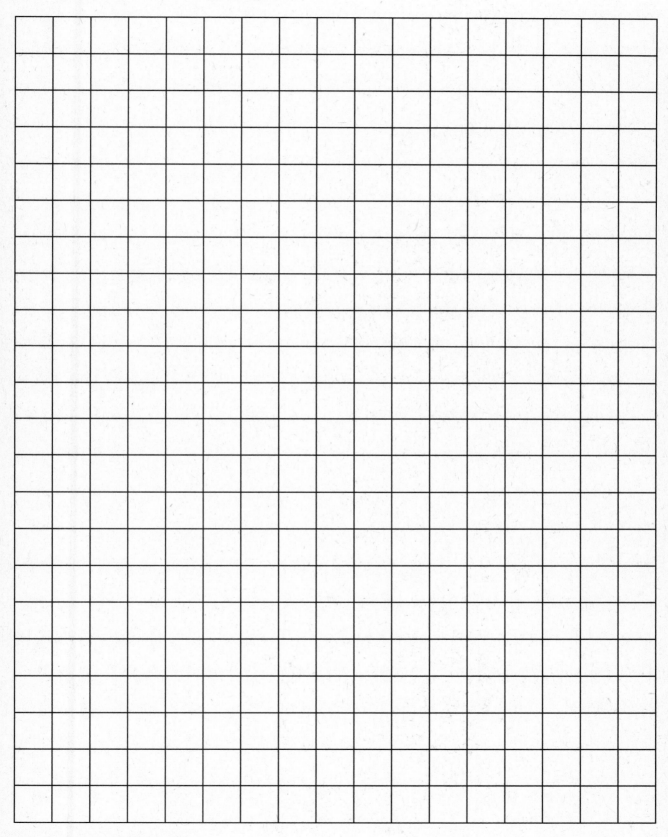

11 | Square Centimeter Grid
For use with Investigation 11

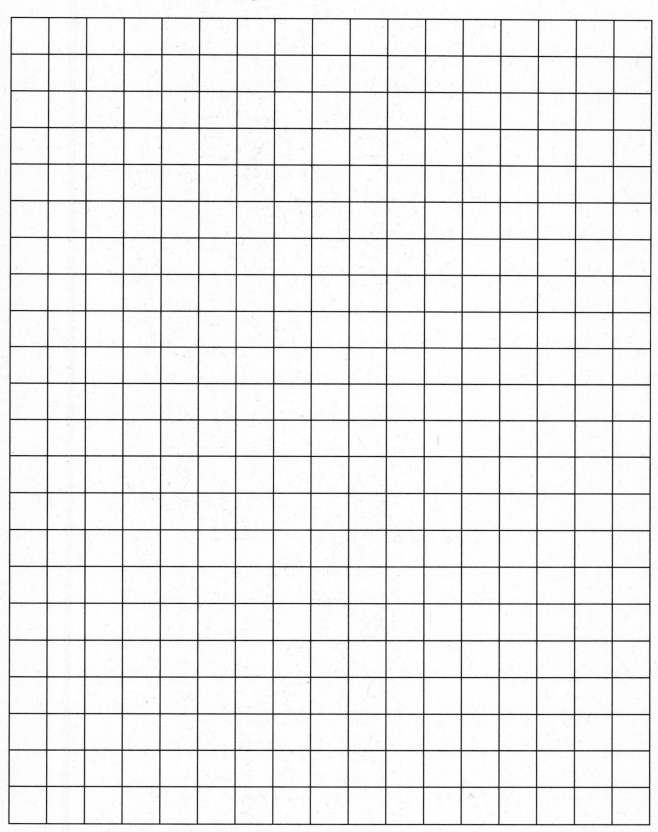

V

$+ - \times \div$ Algebraic Terms
For use with Lesson 111

Name _____

Time _____

Simplify.

$6x + 2x =$	$6x - 2x =$	$(6x)(2x) =$	$\dfrac{6x}{2x} =$
$6xy + 2xy =$	$6xy - 2xy =$	$6xy(2xy) =$	$\dfrac{6xy}{2xy} =$
$x + y + x =$	$x + y - x =$	$(x)(y)(-x) =$	$\dfrac{xy}{x} =$
$3x + x + 3 =$	$3x - x - 3 =$	$(3x)(-x)(-3) =$	$\dfrac{(2x)(8xy)}{4y} =$

$3x + 2y + x - y =$	$5xy - 2x + xy - x =$

Multiplying and Dividing in Scientific Notation

For use with Lesson 112

Name _____

Time _____

Simplify each expression. Write each answer in scientific notation.

$(1 \times 10^6)(1 \times 10^6) =$	$(3 \times 10^3)(3 \times 10^3) =$	$(4 \times 10^{-5})(2 \times 10^{-6}) =$
$(5 \times 10^5)(5 \times 10^5) =$	$(6 \times 10^{-3})(7 \times 10^{-4}) =$	$(3 \times 10^6)(2 \times 10^{-4}) =$
$(9 \times 10^{-6})(2 \times 10^2) =$	$(5 \times 10^8)(4 \times 10^{-2}) =$	$(2.5 \times 10^{-6})(4 \times 10^{-4}) =$
$\dfrac{8 \times 10^8}{2 \times 10^2} =$	$\dfrac{5 \times 10^6}{2 \times 10^3} =$	$\dfrac{9 \times 10^3}{3 \times 10^8} =$
$\dfrac{7.5 \times 10^3}{2.5 \times 10^6} =$	$\dfrac{2 \times 10^6}{4 \times 10^2} =$	$\dfrac{1 \times 10^3}{4 \times 10^8} =$
$\dfrac{6 \times 10^4}{2 \times 10^{-4}} =$	$\dfrac{8 \times 10^{-8}}{2 \times 10^{-2}} =$	$\dfrac{2.5 \times 10^{-4}}{5 \times 10^{-8}} =$

V

+ − × ÷ Algebraic Terms

For use with Lesson 113

Name _____

Time _____

Simplify.

$6x + 2x =$	$6x - 2x =$	$(6x)(2x) =$	$\dfrac{6x}{2x} =$
$6xy + 2xy =$	$6xy - 2xy =$	$6xy(2xy) =$	$\dfrac{6xy}{2xy} =$
$x + y + x =$	$x + y - x =$	$(x)(y)(-x) =$	$\dfrac{xy}{x} =$
$3x + x + 3 =$	$3x - x - 3 =$	$(3x)(-x)(-3) =$	$\dfrac{(2x)(8xy)}{4y} =$
$3x + 2y + x - y =$		$5xy - 2x + xy - x =$	

Multiplying and Dividing in Scientific Notation

For use with Lesson 114

Name _____

Time _____

Simplify each expression. Write each answer in scientific notation.

$(1 \times 10^6)(1 \times 10^6) =$	$(3 \times 10^3)(3 \times 10^3) =$	$(4 \times 10^{-5})(2 \times 10^{-6}) =$
$(5 \times 10^5)(5 \times 10^5) =$	$(6 \times 10^{-3})(7 \times 10^{-4}) =$	$(3 \times 10^6)(2 \times 10^{-4}) =$
$(9 \times 10^{-6})(2 \times 10^2) =$	$(5 \times 10^8)(4 \times 10^{-2}) =$	$(2.5 \times 10^{-6})(4 \times 10^{-4}) =$
$\dfrac{8 \times 10^8}{2 \times 10^2} =$	$\dfrac{5 \times 10^6}{2 \times 10^3} =$	$\dfrac{9 \times 10^3}{3 \times 10^8} =$
$\dfrac{7.5 \times 10^3}{2.5 \times 10^6} =$	$\dfrac{2 \times 10^6}{4 \times 10^2} =$	$\dfrac{1 \times 10^3}{4 \times 10^8} =$
$\dfrac{6 \times 10^4}{2 \times 10^{-4}} =$	$\dfrac{8 \times 10^{-8}}{2 \times 10^{-2}} =$	$\dfrac{2.5 \times 10^{-4}}{5 \times 10^{-8}} =$

V   **+ − × ÷ Algebraic Terms**

For use with Lesson 115

Name _____

Time _____

Simplify.

$6x + 2x =$	$6x - 2x =$	$(6x)(2x) =$	$\dfrac{6x}{2x} =$
$6xy + 2xy =$	$6xy - 2xy =$	$6xy(2xy) =$	$\dfrac{6xy}{2xy} =$
$x + y + x =$	$x + y - x =$	$(x)(y)(-x) =$	$\dfrac{xy}{x} =$
$3x + x + 3 =$	$3x - x - 3 =$	$(3x)(-x)(-3) =$	$\dfrac{(2x)(8xy)}{4y} =$
$3x + 2y + x - y =$		$5xy - 2x + xy - x =$	

$+ - \times \div$ **Algebraic Terms**

For use with Test 22

Name _____

Time _____

Simplify.

$6x + 2x =$	$6x - 2x =$	$(6x)(2x) =$	$\dfrac{6x}{2x} =$
$6xy + 2xy =$	$6xy - 2xy =$	$6xy(2xy) =$	$\dfrac{6xy}{2xy} =$
$x + y + x =$	$x + y - x =$	$(x)(y)(-x) =$	$\dfrac{xy}{x} =$
$3x + x + 3 =$	$3x - x - 3 =$	$(3x)(-x)(-3) =$	$\dfrac{(2x)(8xy)}{4y} =$
$3x + 2y + x - y =$		$5xy - 2x + xy - x =$	

Multiplying and Dividing in Scientific Notation

For use with Lesson 116

Name _____

Time _____

Simplify each expression. Write each answer in scientific notation.

$(1 \times 10^6)(1 \times 10^6) =$	$(3 \times 10^3)(3 \times 10^3) =$	$(4 \times 10^{-5})(2 \times 10^{-6}) =$
$(5 \times 10^5)(5 \times 10^5) =$	$(6 \times 10^{-3})(7 \times 10^{-4}) =$	$(3 \times 10^6)(2 \times 10^{-4}) =$
$(9 \times 10^{-6})(2 \times 10^2) =$	$(5 \times 10^8)(4 \times 10^{-2}) =$	$(2.5 \times 10^{-6})(4 \times 10^{-4}) =$
$\dfrac{8 \times 10^8}{2 \times 10^2} =$	$\dfrac{5 \times 10^6}{2 \times 10^3} =$	$\dfrac{9 \times 10^3}{3 \times 10^8} =$
$\dfrac{7.5 \times 10^3}{2.5 \times 10^6} =$	$\dfrac{2 \times 10^6}{4 \times 10^2} =$	$\dfrac{1 \times 10^3}{4 \times 10^8} =$
$\dfrac{6 \times 10^4}{2 \times 10^{-4}} =$	$\dfrac{8 \times 10^{-8}}{2 \times 10^{-2}} =$	$\dfrac{2.5 \times 10^{-4}}{5 \times 10^{-8}} =$

V

$+ - \times \div$ **Algebraic Terms**

For use with Lesson 117

Name _____

Time _____

Simplify.

$6x + 2x =$	$6x - 2x =$	$(6x)(2x) =$	$\dfrac{6x}{2x} =$
$6xy + 2xy =$	$6xy - 2xy =$	$6xy(2xy) =$	$\dfrac{6xy}{2xy} =$
$x + y + x =$	$x + y - x =$	$(x)(y)(-x) =$	$\dfrac{xy}{x} =$
$3x + x + 3 =$	$3x - x - 3 =$	$(3x)(-x)(-3) =$	$\dfrac{(2x)(8xy)}{4y} =$
$3x + 2y + x - y =$		$5xy - 2x + xy - x =$	

Saxon Math 8/7—Homeschool

Multiplying and Dividing in Scientific Notation

For use with Lesson 118

Name _____

Time _____

Simplify each expression. Write each answer in scientific notation.

$(1 \times 10^6)(1 \times 10^6) =$	$(3 \times 10^3)(3 \times 10^3) =$	$(4 \times 10^{-5})(2 \times 10^{-6}) =$
$(5 \times 10^5)(5 \times 10^5) =$	$(6 \times 10^{-3})(7 \times 10^{-4}) =$	$(3 \times 10^6)(2 \times 10^{-4}) =$
$(9 \times 10^{-6})(2 \times 10^2) =$	$(5 \times 10^8)(4 \times 10^{-2}) =$	$(2.5 \times 10^{-6})(4 \times 10^{-4}) =$
$\dfrac{8 \times 10^8}{2 \times 10^2} =$	$\dfrac{5 \times 10^6}{2 \times 10^3} =$	$\dfrac{9 \times 10^3}{3 \times 10^8} =$
$\dfrac{7.5 \times 10^3}{2.5 \times 10^6} =$	$\dfrac{2 \times 10^6}{4 \times 10^2} =$	$\dfrac{1 \times 10^3}{4 \times 10^8} =$
$\dfrac{6 \times 10^4}{2 \times 10^{-4}} =$	$\dfrac{8 \times 10^{-8}}{2 \times 10^{-2}} =$	$\dfrac{2.5 \times 10^{-4}}{5 \times 10^{-8}} =$

V $+ - \times \div$ **Algebraic Terms**
For use with Lesson 119

Name _____

Time _____

Simplify.

$6x + 2x =$	$6x - 2x =$	$(6x)(2x) =$	$\dfrac{6x}{2x} =$
$6xy + 2xy =$	$6xy - 2xy =$	$6xy(2xy) =$	$\dfrac{6xy}{2xy} =$
$x + y + x =$	$x + y - x =$	$(x)(y)(-x) =$	$\dfrac{xy}{x} =$
$3x + x + 3 =$	$3x - x - 3 =$	$(3x)(-x)(-3) =$	$\dfrac{(2x)(8xy)}{4y} =$
$3x + 2y + x - y =$		$5xy - 2x + xy - x =$	

Multiplying and Dividing in Scientific Notation

For use with Lesson 120

Name _____

Time _____

Simplify each expression. Write each answer in scientific notation.

$(1 \times 10^6)(1 \times 10^6) =$	$(3 \times 10^3)(3 \times 10^3) =$	$(4 \times 10^{-5})(2 \times 10^{-6}) =$
$(5 \times 10^5)(5 \times 10^5) =$	$(6 \times 10^{-3})(7 \times 10^{-4}) =$	$(3 \times 10^6)(2 \times 10^{-4}) =$
$(9 \times 10^{-6})(2 \times 10^2) =$	$(5 \times 10^8)(4 \times 10^{-2}) =$	$(2.5 \times 10^{-6})(4 \times 10^{-4}) =$
$\dfrac{8 \times 10^8}{2 \times 10^2} =$	$\dfrac{5 \times 10^6}{2 \times 10^3} =$	$\dfrac{9 \times 10^3}{3 \times 10^8} =$
$\dfrac{7.5 \times 10^3}{2.5 \times 10^6} =$	$\dfrac{2 \times 10^6}{4 \times 10^2} =$	$\dfrac{1 \times 10^3}{4 \times 10^8} =$
$\dfrac{6 \times 10^4}{2 \times 10^{-4}} =$	$\dfrac{8 \times 10^{-8}}{2 \times 10^{-2}} =$	$\dfrac{2.5 \times 10^{-4}}{5 \times 10^{-8}} =$

Multiplying and Dividing in Scientific Notation

For use with Test 23

Name _____

Time _____

Simplify each expression. Write each answer in scientific notation.

$(1 \times 10^6)(1 \times 10^6) =$	$(3 \times 10^3)(3 \times 10^3) =$	$(4 \times 10^{-5})(2 \times 10^{-6}) =$
$(5 \times 10^5)(5 \times 10^5) =$	$(6 \times 10^{-3})(7 \times 10^{-4}) =$	$(3 \times 10^6)(2 \times 10^{-4}) =$
$(9 \times 10^{-6})(2 \times 10^2) =$	$(5 \times 10^8)(4 \times 10^{-2}) =$	$(2.5 \times 10^{-6})(4 \times 10^{-4}) =$
$\dfrac{8 \times 10^8}{2 \times 10^2} =$	$\dfrac{5 \times 10^6}{2 \times 10^3} =$	$\dfrac{9 \times 10^3}{3 \times 10^8} =$
$\dfrac{7.5 \times 10^3}{2.5 \times 10^6} =$	$\dfrac{2 \times 10^6}{4 \times 10^2} =$	$\dfrac{1 \times 10^3}{4 \times 10^8} =$
$\dfrac{6 \times 10^4}{2 \times 10^{-4}} =$	$\dfrac{8 \times 10^{-8}}{2 \times 10^{-2}} =$	$\dfrac{2.5 \times 10^{-4}}{5 \times 10^{-8}} =$

12 | Pythagorean Puzzle

For use with Investigation 12

Carefully cut out right triangle *ABC* and the squares drawn on the legs of the triangle. Then cut each square into four parts and reassemble all eight parts to form a square on the hypotenuse of triangle *ABC*. Completing this puzzle illustrates that the sum of the areas of the squares drawn on the legs of a right triangle equals the area of a square drawn on the hypotenuse.

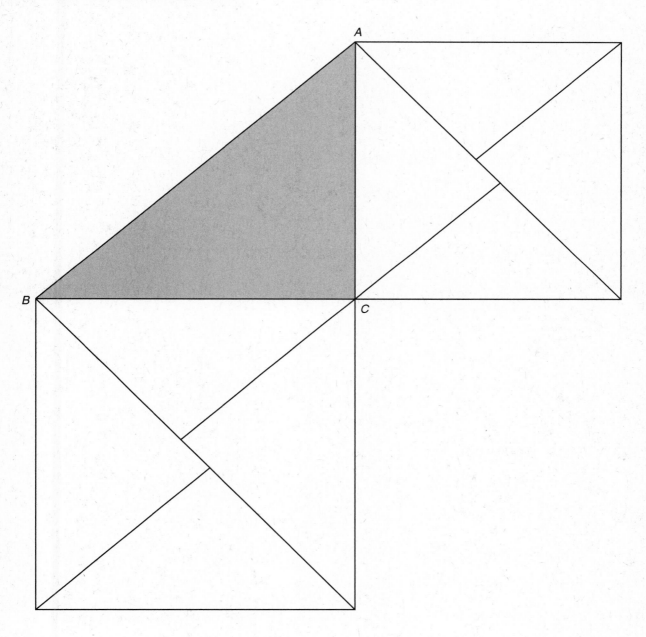

Tests

A test should be given after every fifth lesson, beginning after Lesson 10. The testing schedule is explained in greater detail on the back of this page.

On test days, allow five minutes for your student to take the Facts Practice Test indicated at the top of the test. Then administer the cumulative test specified by the Testing Schedule. You might wish also to provide your student with a photocopy of Recording Form E. This form is designed to provide an organized space for your student to show his or her work. *Note:* The textbook should not be used during the test.

Solutions to the test problems are located in the *Saxon Math 8/7—Homeschool Solutions Manual.* For detailed information on appropriate test-grading strategies, please refer to the preface in the *Saxon Math 8/7—Homeschool* textbook.

Testing Schedule

Test to be administered	Covers material through	Give after
Test 1	Lesson 5	Lesson 10
Test 2	Lesson 10	Lesson 15
Test 3	Lesson 15	Lesson 20
Test 4	Lesson 20	Lesson 25
Test 5	Lesson 25	Lesson 30
Test 6	Lesson 30	Lesson 35
Test 7	Lesson 35	Lesson 40
Test 8	Lesson 40	Lesson 45
Test 9	Lesson 45	Lesson 50
Test 10	Lesson 50	Lesson 55
Test 11	Lesson 55	Lesson 60
Test 12	Lesson 60	Lesson 65
Test 13	Lesson 65	Lesson 70
Test 14	Lesson 70	Lesson 75
Test 15	Lesson 75	Lesson 80
Test 16	Lesson 80	Lesson 85
Test 17	Lesson 85	Lesson 90
Test 18	Lesson 90	Lesson 95
Test 19	Lesson 95	Lesson 100
Test 20	Lesson 100	Lesson 105
Test 21	Lesson 105	Lesson 110
Test 22	Lesson 110	Lesson 115
Test 23	Lesson 115	Lesson 120

1

1. When the product of 12 and 60 is divided by the sum of 12 and 36, what is the quotient?
(1)

2. Use the numbers 4 and 5 to illustrate the commutative property of multiplication.
(2)

3. Use digits and symbols to write "Negative five is less than positive three."
(4)

4. Use words to write 43080070.
(5)

5. Find the next three numbers in this sequence.
(2)

$$1, 4, 9, 16, \ldots$$

6. Write 130,500 in expanded notation.
(5)

7. Replace the circle with the proper comparison symbol.
(4)

$$-6 \bigcirc -8$$

8. If $a = 14$ and $b = 4$, then what does ab equal?
(1)

9. Use digits to write three million, forty thousand, seven hundred.
(5)

Find each missing number:

10. $t + \$5.50 = \$12.00$
(3)

11. $b - 9630 = 2480$
(3)

12. $7f = \$51.80$
(3)

13. $6048 - y = 2532$
(3)

14. $15p = 1275$
(3)

15. $18,400 - c = 7520$
(3)

Simplify:

16. $9 \cdot 22 \cdot 25$
(1)

17. $1000 - (720 - 38)$
(2)

18. $6\overline{)38,154}$
(1)

19. $170(18)$
(1)

20. $\dfrac{\$41.30}{10}$
(1)

2

*Also take Facts Practice Test B
(30 Equations).*

Name _____

1. Three dimes is
(8)
  (a) what fraction of a dollar?   (b) what percent of a dollar?

2. In this triangle, which segment is perpendicular to $\overline{BC}$?
(7)

3. How many $\dfrac{5}{8}$'s are in 1?
(9)

4. Write $9\dfrac{5}{8}$ as an improper fraction.
(10)

5. (a) Arrange these numbers in order from least to greatest:
(4)
$$8, -8, \frac{1}{8}, 0$$

  (b) Which of the numbers in part (a) is not an integer?

6. Use the numbers 2, 6, and 8 to illustrate the associative property of addition.
(2)

7. (a) What fraction of the rectangle is shaded?
(8)
  (b) What fraction of the rectangle is not shaded?

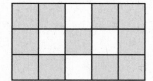

8. Subtract fifty-six million from three hundred million, and use words to write the difference.
(5)

9. (a) List the factors of 28.
(6)
  (b) List the factors of 42.
  (c) Which numbers are factors of both 28 and 42?
  (d) What is the greatest common factor of 28 and 42?

10. Use digits and symbols to write "The product of four and one is less than the sum of four and one."
(4)

Find each missing number:

11. $3955 + c = 7000$
(3)

12. $g - \$4.85 = \$6.45$
(3)

13. $34p = 782$
(3)

Simplify:

14. $\dfrac{1}{11} + \dfrac{2}{11}$
(9)

15. $\dfrac{7}{13} - \dfrac{6}{13}$
(9)

16. $\dfrac{4}{7} \times \dfrac{4}{9}$
(9)

17. $8\overline{)34{,}594}$
(1)

18. $90(\$7.44)$
(1)

19. $\dfrac{3}{4} \cdot \dfrac{1}{8} \cdot \dfrac{5}{7}$
(9)

20. Describe each figure as a line, ray, or segment. Then use a symbol and letters to name each figure.
(7)
  (a)   (b) *L*   (c)

3

Name _____

1. The population of Hinchton in 2002 was 22,374. In 1997, the population was only 14,998. How much did
(12) the population increase between 1997 and 2002?

2. Levi received a shipment of 13 boxes of T-shirts. Each box contained 18 T-shirts. How many T-shirts
(13) were in the shipment?

3. The product of 8 and 4 is how much greater than the sum of 8 and 4?
(1, 12)

4. Alicia spent $6.75 for the ticket, $3.95 for popcorn, and 95¢ for a drink. How much did she spend in all?
(11)

5. How many years were there from 1556 to 1728?
(12)

6. If 29% of the campers wore jeans, what percent of the campers did not wear jeans?
(14)

7. Draw and shade circles to show that $1\dfrac{3}{8} = \dfrac{11}{8}$.
(10)

8. Complete each equivalent fraction.
(15)

(a) $\dfrac{2}{3} = \dfrac{?}{48}$ (b) $\dfrac{5}{8} = \dfrac{?}{48}$

9. Find a fraction equal to $\frac{1}{5}$ that has a denominator of 10. Then subtract $\frac{1}{10}$ from that fraction. What is the
(15) difference?

10. (a) List the factors of 27.
(6)
    (b) List the factors of 63.
    (c) What is the greatest common factor of 27 and 63?

11. Name three segments in this figure in order of length from shortest to longest.
(7)

12. What mixed number is represented by point *A* on this number line?
(8)

Simplify:

13. $\dfrac{11}{22} + \dfrac{13}{22}$
(10)

14. $\dfrac{3}{4} \cdot \dfrac{8}{5}$
(15)

15. $5\overline{)38{,}519}$
(1)

16. $\dfrac{1060}{20}$
(1)

17. $\begin{array}{r} 122 \\ \times\ \ 84 \end{array}$
(1)

18. $(4 + 3)(3)$
(2)

Find each missing number:

19. $14t = 1820$
(3)

20. $\$20.00 - w = \$4.52$
(3)

Also take Facts Practice Test D
(40 Fractions to Reduce).

Name _____

1. Great-Grandma celebrated her eighty-fourth birthday in 2002. In what year was she born?
(12)

2. The farmer harvested 5000 bushels of barley from 40 acres. The crop produced an average of how many
(13) bushels of barley for each acre?

3. Two feet is what percent of one yard?
(8, 16)

4. At the beginning of the day there were 689 tickets available for the ballgame. By the end of the day, all
(11) but 39 tickets were sold. How many tickets were sold that day?

5. Seventy-one million is how much less than one billion? Use words to write the answer.
(5, 12)

6. (a) Compare: $\dfrac{5}{16} + \left(\dfrac{3}{16} + \dfrac{7}{16}\right) \bigcirc \left(\dfrac{5}{16} + \dfrac{3}{16}\right) + \dfrac{7}{16}$
(2, 9)

    (b) What property is illustrated by this comparison?

7. Use digits and symbols to write "Six minus nine equals negative three."
(4)

8. (a) Find the perimeter of this rectangle.
(19, 20)
    (b) Find the area of this rectangle.

                                           5 cm

                     12 cm

9. Reduce:
(15)

    (a) $8\dfrac{10}{16}$ (b) $\dfrac{2}{16}$

10. Write $2\dfrac{1}{4}$ as an improper fraction, and multiply the improper fraction by $\dfrac{1}{9}$.
(9, 10)

11. Complete each equivalent fraction.
(15)

    (a) $\dfrac{3}{5} = \dfrac{?}{40}$ (b) $\dfrac{7}{10} = \dfrac{?}{40}$

12. Use a protractor to draw a 43° angle.
(17)

Find each missing number:

13. $3446 - n = 1428$ **14.** $30j = \$55.50$
(3)   (3)

15. Which of the following does not equal $3\dfrac{1}{3}$?
(15)

    A. $3\dfrac{2}{6}$ B. $\dfrac{10}{3}$ C. $3\dfrac{11}{33}$ D. $\dfrac{7}{3}$

Simplify:

16. $\dfrac{3}{8} + \dfrac{3}{8} + \dfrac{3}{8}$ **17.** $\dfrac{11}{12} - \dfrac{5}{12}$ **18.** $\left(\dfrac{1}{4}\right)^2$ **19.** $\sqrt{256}$ **20.** $14(10 + 11)$
(9)   (9)   (9, 20)   (20)   (2)

Also take Facts Practice Test E (Circles).   Name _____

1. Six hundred twenty-one books were packed into 27 boxes. If each box contained the same number of
(13) books, how many books were packed in each box?

2. The Holy Roman Empire lasted from 800 to 1806. How many years did the Holy Roman Empire last?
(12)

3. Mya went to the ball game with $20.00 and returned with $8.24. How much money did Mya spend at the
(11) ball game?

4. Michael was engrossed in his 340-page book. He stopped on page 197 at noon to eat lunch. He stopped on
(12) page 283 to eat dinner. How many pages did Michael read between lunch and dinner?

5. Diagram this statement. Then answer the questions that follow.
(22)
Two ninths of the 81 fish in the tank were guppies.

(a) How many of the fish in the tank were guppies?

(b) How many of the fish in the tank were not guppies?

6. Use a compass and straightedge to inscribe a regular hexagon in a circle.
(Inv. 2)

7. Write the prime factorization of 360.
(21)

8. Simplify:
(15)
(a) $\dfrac{105}{9}$ (b) $3\dfrac{8}{5}$ (c) $\dfrac{640}{780}$

9. Write the reciprocal of each number.
(9)
(a) $\dfrac{3}{7}$ (b) $9\dfrac{2}{3}$ (c) 6

10. Complete each equivalent fraction.
(15)
(a) $\dfrac{3}{4} = \dfrac{?}{28}$ (b) $\dfrac{5}{7} = \dfrac{?}{28}$

Solve:

11. $610 = 900 - c$ **12.** $g - 57 = 56$ **13.** $15h = 465$
(3)   (3)   (3)

Simplify:

14. $8 - 1\dfrac{2}{3}$ **15.** $6\dfrac{4}{7} + 5\dfrac{6}{7}$ **16.** $4\dfrac{1}{6} - 2\dfrac{5}{6}$
(23)   (10)   (23)

17. $\dfrac{2}{5} \cdot \dfrac{5}{8} \cdot \dfrac{8}{9}$ **18.** $\dfrac{3}{4} \div \dfrac{2}{3}$ **19.** $10^2 - \sqrt{36}$
(24)   (25)   (20)

20. Refer to rectangle *ABCD* to answer questions (a)
(7, 20) and (b).

(a) Which side of the rectangle is parallel to $\overline{BC}$?

(b) If *AB* is 28 mm and *BC* is 14 mm, what is the
area of the rectangle?

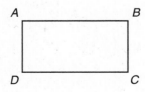

Also take Facts Practice Test F   Name _____
(Lines, Angles, Polygons).

1. The 5 starters on the basketball team were tall. Their heights were 82 inches, 74 inches, 78 inches,
(28) 80 inches, and 76 inches. What was the average height of the 5 starters?

2. Linda bought 9 pounds of peaches for $0.87 per pound and paid for them with a ten-dollar bill. How much
(28) should she get back in change?

3. On the first day of their 1634-mile trip, the Zabrockys drove 276 miles. How many more miles do they
(11) have to drive until they complete their trip?

4. The coordinates of three vertices of a rectangle are (1, –2), (7, –2), and (7, 3).
(Inv. 3)
    (a) What are the coordinates of the fourth vertex?

    (b) What is the area of the rectangle?

5. Diagram this statement. Then answer the questions that follow.
(22)
                The Chins completed 25% of their 2540-mile trip the first day.

    (a) How many miles did they travel the first day?

    (b) How many miles of their trip did they still have to travel?

6. If the perimeter of a square is 7 feet, how many inches long is each side of the square?
(19)

7. Rewrite $\frac{3}{5}$ and $\frac{2}{3}$ so that they have common denominators. Then add the fractions and simplify the sum.
(30)

8. (a) Round 68,261 to the nearest thousand.   (b) Round 68,261 to the nearest hundred.
(29)

9. Estimate the quotient when 38,472 is divided by 41.
(29)

10. Reduce: $\dfrac{160}{240}$
(24)

11. Compare: $\dfrac{7}{8}$ ◯ $\dfrac{8}{7}$
(30)

12. Find the least common multiple (LCM) of 8 and 10.
(27)

13. The figure shows a circle with the center at Q.
(7, Inv. 2)
    (a) Which chord is a diameter?

    (b) Which inscribed angle appears to be a right angle?

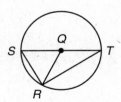

14. (a) Write the prime factorization of 324.   (b) Find $\sqrt{324}$.
(21)

Solve:

15. $9w = 4 \cdot 36$ **16.** $287 + r = 971$ **17.** $94 - h = 49$
(3)   *(3)*   *(3)*

Simplify:

18. $\dfrac{3}{5} + \dfrac{1}{2}$ **19.** $\left(\dfrac{2}{3} \cdot \dfrac{5}{6}\right) - \dfrac{2}{5}$ **20.** $5\dfrac{1}{3} \div 1\dfrac{7}{9}$
(30)   *(30)*   *(26)*

1. In the first four months of the year the Morrisons' electric bills were $113.96, $99.21, $93.20, and
(28) $128.95. What was the Morrisons' average electricity bill during the first four months of the year?

2. The price was reduced from nine thousand, five hundred twenty-one dollars to three thousand, two hundred
(12) sixty-seven dollars. By how much was the price reduced?

3. A one-year subscription to the monthly magazine costs $41.40. The regular newsstand price is $4.25 per
(28) issue. How much is saved per issue by paying the subscription price?

4. Ivan ran one lap in 1 minute 2 seconds. Donovan ran one lap 7 seconds faster than Ivan. How many
(28) seconds did it take Donovan to run one lap?

5. The perimeter of the square equals the perimeter of
(19) the regular pentagon. Each side of the pentagon is
24 cm. How long is each side of the square?

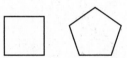

6. Diagram this statement. Then answer the questions that follow.
(22)

> *Two fifths of the 45 fish in the tank were guppies.*

   (a) How many of the fish were guppies?   (b) How many of the fish were not guppies?

7. Find the least common multiple (LCM) of 5, 8, and 10.
(27)

8. Round 1832.2243 (a) to the nearest hundredth and (b) to the nearest hundred.
(33)

9. (a) What fraction of this square is shaded?
(8)   (b) What percent of this square is shaded?

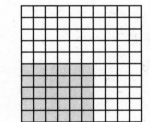

10. What decimal number is halfway between 59 and 60?
(34)

11. The coordinates of three vertices of a rectangle are
(Inv. 3) (−5, −1), (4, −1), and (4, −8).

   (a) What are the coordinates of the fourth vertex?

   (b) What is the area of the rectangle?

12. What decimal number names the point marked with an arrow on this number line?
(34)

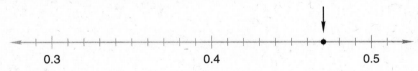

13. Solve: $14x = 7 \cdot 10$ **14.** Use words to write 237.208.
(3)   (31)

Simplify:

15. 2.5×2.5 **16.** $2.73 \div 3$ **17.** $8.5 + 1.83 + 15$
(35)   (35)   (35)

18. $31.74 - 2.146$ **19.** $4\frac{1}{3} - \left(\frac{1}{9} \cdot \frac{3}{4} \right)$ **20.** $\left(3\frac{1}{3} + 1\frac{1}{4} \right) \div \left(6 - 4\frac{1}{6} \right)$
(35)   (9, 30)   (26, 30)

8

*Also take Facts Practice Test H
(Measurement Facts).*

Name _____

1. The bag contained only red marbles and white marbles. If the ratio of red marbles to white marbles was
(36) 7 to 5, what fraction of the marbles were white?

2. Mia ran 5 laps of the track in 6 minutes 50 seconds.
(28)
(a) How many seconds did it take Mia to run 5 laps?

(b) What was the average number of seconds it took Mia to run each lap?

3. Lupe's car averages 27 miles per gallon of gas. At that rate, how far would it go on 27 gallons?
(46)

4. Diagram this statement. Then answer the questions that follow.
(22)
Seventy-five percent of the 116 adults in the McKenna clan were 5 feet tall or taller.

(a) How many of the adults were less than 5 feet tall?

(b) How many of the adults were 5 feet tall or taller?

Refer to the figure at right for problems 5 and 6. All angles are right angles.

5. What is the perimeter of this polygon?
(19)

6. What is the area of this polygon?
(37)

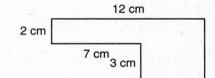

7. *AB* is 56 mm. *CD* is 14 mm. *AD* is 98 mm. Find *BC*.
(7)

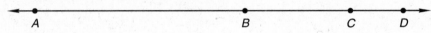

8. The length of segment *AB* in problem 7 is 56 mm. What is the length of segment *AB* in centimeters?
(32)

9. Round 0.870547 to the nearest thousandth.
(33)

10. If two angles of a triangle each measure 40°, then what is the measure of the third angle of the triangle?
(40)

11. What decimal number names point *C* on this number line?
(34)

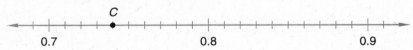

12. Write forty-eight and nine hundredths (a) as a decimal number and (b) as a mixed number.
(31)

13. What is the area of this triangle?
(37)

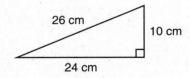

Simplify:

14. 0.37(0.04)
(35)

15. 0.156 ÷ 6
(35)

16. $8\frac{1}{6} - 4\frac{2}{3}$
(23)

17. $2\frac{1}{4} \cdot 2\frac{2}{3}$
(26)

18. $8 \div 2\frac{2}{3}$
(26)

19. Solve: $\frac{8}{12} = \frac{w}{15}$
(39)

20. Solve: $m + 0.34 = 2.04$
(3, 35)

9

Also take Facts Practice Test I *(Proportions).*

Name _____

1. What is the probability of rolling a prime number with one roll of a die?
(36)

2. Amber's test scores were 100, 98, 91, 84, 93, 88, 97, 91, 87, and 91.
(28, Inv. 4)
  (a) Find the mean of her scores.   (b) Find the median of her scores.

3. Evaluate: $a(b + c)$ if $a = 0.6$, $b = 4.1$, and $c = 0.7$
(41)

4. Refer to this election tally sheet to answer questions
(38)
  (a) and (b).
    (a) The first-place candidate received how many
        more votes than the fourth-place candidate?
    (b) What fraction of the votes did Judy receive?

Vote Totals

Judy	𝍸𝍸 𝍸𝍸 𝍸𝍸 I
Carlos	𝍸𝍸 𝍸𝍸 IIII
Yolanda	𝍸𝍸 𝍸𝍸 𝍸𝍸 𝍸𝍸 II
Khanh	𝍸𝍸 𝍸𝍸 𝍸𝍸 III

5. Find the area of a triangle whose vertices have the coordinates (–1, 7), (–1, 2), and (5, 2).
(Inv. 3, 37)

6. Read this statement. Then answer the questions that follow.
(22, 36)

> *Five twelfths of those who rode the Giant Gyro at the fair were euphoric.*
> *All the rest were vertiginous.*

  (a) What fraction of those who rode the ride were vertiginous?
  (b) What was the ratio of euphoric to vertiginous riders?

7. The perimeter of the rectangle is 64 cm. What is
(19) the length of the rectangle?

14 cm

8. Find m∠a, m∠b, and m∠c in this figure.
(40)

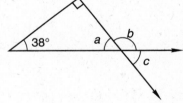

38°

9. Write 375% as a decimal number.
(43)

10. Write $\frac{7}{5}$ as a decimal number.
(43)

11. Round $67.\overline{24}$ to four decimal places.
(42)

12. Divide 3.2 by 9 and write the quotient with a bar over the repetend.
(42)

Solve:

13. $\dfrac{14}{6} = \dfrac{7}{m}$
(39)

14. $8 = 3.14 + x$
(3, 35)

15. $0.072 = 1 - z$
(3, 35)

Simplify:

16. $5\dfrac{3}{4} + \dfrac{3}{5} + 3\dfrac{1}{2}$
(30)

17. $4\dfrac{1}{6} - \left(3 - 1\dfrac{1}{4}\right)$
(23)

18. $3\dfrac{1}{5} \cdot 4\dfrac{3}{8} \cdot 2$
(26)

19. $8 \div 10\dfrac{2}{3}$
(26)

20. $1.68 \div 0.6$
(35)

10

Also take Facts Practice Test J
(+ − × ÷ Decimals).

Name _____

1. What is the total cost of a $19.98 item plus 7% sales tax?
(46)

2. Brand X costs $2.73 for 13 ounces. Brand Y costs 6¢ more per ounce. What is the cost of 16 ounces of
(28) Brand Y?

3. The ratio of pansies to petunias in the garden was 17 to 8. What was the ratio of petunias to pansies?
(36)

4. During the month of February, Hannah's weekly grocery bills were $119.97, $98.58, $99.18, and
(28) $105.15. Find Hannah's average weekly grocery bill in February to the nearest dollar.

5. Three and one hundredths is how much less than three and five tenths? Write the answer in words.
(31, 35)

6. Diagram this statement. Then answer the questions that follow.
(22)
$$\textit{Sixty percent of the 80 boats at the dock were for sale.}$$

    (a) What fraction of the boats were not for sale?

    (b) How many boats were not for sale?

7. Find the area of this figure.
(37)

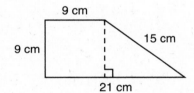

8. Write 16% as a reduced fraction.
(43)

9. Divide 3.8 by 11 and write the answer with a bar over the repetend.
(42)

10. Reduce: $\dfrac{630}{810}$
(15)

11. If the perimeter of a square is 36 inches, what is its area?
(19, 20)

Solve:

12. $\dfrac{45}{72} = \dfrac{25}{f}$
(39)

13. $4j = 9.6$
(3, 35)

14. $9 - y = 1.83$
(3, 35)

Simplify:

15. $7^2 - 3^3$
(20)

16. $\begin{array}{r} 2 \text{ hr } 47 \text{ min } 50 \text{ s} \\ + 3 \text{ hr } 34 \text{ min } 45 \text{ s} \\ \hline \end{array}$
(49)

17. $15\dfrac{3}{12} - 7\dfrac{1}{8}$
(30)

18. $6\dfrac{1}{4} \div 3\dfrac{1}{8}$
(26)

19. 0.185×10^4
(47)

20. $0.2106 \div 0.04$
(45)

11

Also take Facts Practice Test K
(Powers and Roots).

Name _____

1. The ratio of yachts to sloops in the bay was 3 to 7. If there were 54 yachts in the bay, how many sloops were there?
(54)

2. The average of four numbers is 91. If three of the numbers are 89, 84, and 92, what is the fourth number?
(55)

3. A quart of milk costs 89¢. A case of 12 quarts costs $9.36. How much is saved per quart by buying the milk by the case?
(28)

4. Segment *AB* is how much longer than segment *BC*?
(8)

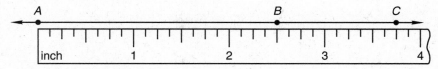

5. Diagram this statement. Then answer the questions that follow.
(22)

Four fifths of the 75 classic cars were hot rods.

(a) How many of the classic cars were hot rods?   (b) What percent of the classic cars were hot rods?

6. (a) Write thirty-four billion in scientific notation.   (b) Write 6.51×10^8 in standard form.
(51)

7. Compare: $1.92 + 0.3 \bigcirc 7 - 5.08$
(33, 35)

8. Use a unit multiplier to convert 710 mm to cm.
(50)

9. Complete this table.
(48)

Fraction	Decimal	Percent
(a)	(b)	450%
$\frac{7}{10}$	(c)	(d)

10. Evaluate: $ab - bc$ if $a = 5$, $b = 4$, and $c = 3$
(52)

Refer to the figure at right for problems 11 and 12. Dimensions are in inches. All angles are right angles.

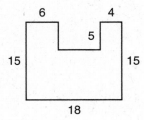

11. What is the area of the figure?
(37)

12. What is the perimeter of the figure?
(19)

13. Solve: $4.36 + w = 10$
(3, 35)

14. Solve: $\dfrac{a}{6} = \dfrac{35}{10}$
(39)

Simplify:

15. $15^2 - 4^3 - 2^4 - \sqrt{225}$
(20)

16. $5 + 5 \cdot 5 - 5 \div 5$
(52)

17. $4\dfrac{3}{4} + 2\dfrac{1}{12} + 1\dfrac{1}{8}$
(30)

18. $4\dfrac{4}{5} \cdot 3\dfrac{1}{8} \cdot 1\dfrac{9}{20}$
(26)

19. $0.8(0.25)(0.04)$
(35)

20. $7.2 \div 0.018$
(45)

1. If a half gallon of milk costs $1.48, what is the cost per pint?
(16)

2. The cookie recipe called for oatmeal and raisins in the ratio of 4 to 1. If 3 cups of oatmeal were called for,
(54) how many cups of raisins were needed?

3. Marcie ran the 400-meter race 3 times. Her fastest time was 51.3 seconds. Her slowest time was
(55) 56.4 seconds. If Marcie's average time was 53.4 seconds, what was her time for the third race?

4. It is $4\frac{1}{2}$ miles to the end of the trail. If Sophia bikes to the end and back in 60 minutes, what is her average
(46) speed in miles per hour?

5. What number is 30% of 50?
(60)

6. Read this statement. Then answer the questions that follow.
(22, 36)

> *Only six tenths of the print area of the newspaper carried news. The rest of*
> *the area was filled with advertisements.*

(a) What percent of the print area was filled with advertisements?

(b) What was the ratio of news area to advertisement area?

7. (a) Write 0.000309 in scientific notation.   (b) Write 4.42×10^{-6} in standard form.
(57)

8. Sketch a quadrilateral with only one pair of parallel sides. What is this type of quadrilateral called?
(Inv. 6)

9. Use a unit multiplier to convert 1050 yards to feet.
(50)

10. Complete this table.
(48)

Fraction	Decimal	Percent
(a)	(b)	18%
$\frac{1}{50}$	(c)	(d)

Refer to the figure at right for problems 11 and 12.
Dimensions are in centimeters. All angles are right
angles.

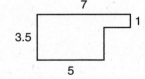

11. What is the perimeter of the figure?
(19)

12. What is the area of the figure?
(37)

Solve:

13. $\dfrac{8}{18} = \dfrac{n}{54}$ **14.** $p + 4.8 = 8$
(39)   (3, 35)

Simplify:

15. $3 + 3 \times 3 - 3 \div 3$ **16.** $10^3 - \sqrt{144} + 3^3 + 5^0$ **17.** $\begin{array}{r} 7 \text{ yd} \\ - 4 \text{ yd } 2 \text{ ft } 9 \text{ in.} \\ \hline \end{array}$
(52)   (20)   (56)

18. $5\frac{2}{5} + \left(4\frac{1}{2} - 2\frac{5}{6}\right)$ **19.** $5\frac{5}{6} \div \left(2\frac{6}{7} \div 4\right)$ **20.** $5.3(0.03)(0.009)$
(30)   (26)   (35)

13

Also take Facts Practice Test M (Metric Conversions).

Name _____

1. What is the total price of a $20,000 car plus 8.5% sales tax?
(46)

2. Emily worked for 9 hours and earned $60.75. How much did she earn per hour?
(46)

3. The ratio of boys to girls in the park was 4 to 5. If 270 children were in the park, how many girls were there?
(65)

4. What is the average of $3\frac{1}{2}$, $4\frac{1}{3}$, 3, and $5\frac{1}{6}$?
(28, 30)

5. What number is 18% of 350?
(60)

6. Diagram this statement. Then answer the questions that follow.
(22, 48)

Joseph gave three fourths of his 228 sports cards to his brother.

   (a) What percent of his sports cards did Joseph give to his brother?

   (b) How many sports cards did Joseph have left?

7. (a) Write 0.0002 in scientific notation.
(57)

   (b) Write 7.1×10^{-2} in standard form.

8. In parallelogram *ABCD*, m∠*C* is 100°. Find m∠*D*.
(61)

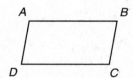

9. Compare: 7.5 kg ◯ 7500 g
(32)

10. Divide 6 by 0.33 and write the answer rounded to the nearest whole number.
(33, 45)

11. Find the sum: $(-2) + (+7) + (-9) + (+3)$
(64)

12. Complete this table.
(48)

Fraction	Decimal	Percent
$\frac{3}{5}$	(a)	(b)
(c)	0.16	(d)

13. Find the area of this parallelogram.
(61)

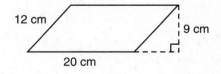

14. Evaluate: $ab + a + b$ if $a = \frac{3}{4}$ and $b = \frac{1}{2}$
(52)

Solve:

15. $\frac{w}{35} = \frac{16}{20}$
(39)

16. $1.6p = 0.256$
(3, 45)

Simplify:

17. $100 - 2[3(6 - 2)]$
(52, 63)

18. $2\frac{3}{4} + \left(4\frac{1}{6} - 3\frac{2}{3}\right)$
(30)

19. $6\frac{7}{8}\left(8 \div 2\frac{3}{4}\right)$
(26)

20. $0.06(0.2)(2.4)$
(35)

1. Simone raced 280 kilometers from Perry to Medford and then idled back. If the round trip took 7 hours, what was Simone's average speed in kilometers per hour?
(46)

2. The ratio of bunnies to squirrels was 3 to 5. If there were 168 bunnies and squirrels in all, how many squirrels were there?
(65)

3. Using a tape measure, Susan found that the circumference of the great redwood was 1200 cm. She estimated that its diameter was 400 cm. Was her estimate a little too large or a little too small? Why?
(66)

4. Almonds were priced at 4 pounds for $11.52.
(53)

(a) What was the price per pound?   (b) How much would 10 pounds of almonds cost?

5. If the product of five tenths and three tenths is subtracted from the sum of two tenths and six tenths, what is the difference?
(35)

6. Diagram this statement. Then answer the questions that follow.
(22, 48)

Three fifths of the baker's 40 cookies were chocolate cookies.

(a) How many of the baker's cookies were chocolate?

(b) What percent of the baker's cookies were not chocolate?

7. (a) A cube has how many faces?
(67, 70)
    (b) What is the volume of this cube?

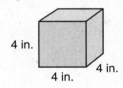

4 in.
4 in.
4 in.

8. Find the circumference of each circle.
(66)

(a)

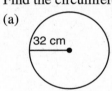

32 cm

Leave π as π.

(b)

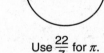

49 mm

Use $\frac{22}{7}$ for π.

9. Write each of these numbers in scientific notation: (a) 14×10^{-7} (b) 14×10^{7}
(69)

Refer to the figure at right for problems 10 and 11. Dimensions are in millimeters.

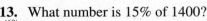
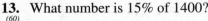

20
15   25   15
25

10. What is the area of the right triangle?
(37, 62)

11. What is the area of the isosceles triangle?
(37, 62)

12. Evaluate: $ab - (a - b)$ if $a = 0.7$ and $b = 0.6$
(52)

13. What number is 15% of 1400?
(60)

14. Complete this table.
(48)

15. Use a unit multiplier to convert 6000 g to kg.
(50)

Solve:

16. $q + 45 = 50.5$
(Inv. 7)

17. $6n = 50$
(Inv. 7)

FRACTION	DECIMAL	PERCENT
$\frac{7}{8}$	(a)	(b)
(c)	(d)	22%

Simplify:

18. $3.2 \times 4\frac{1}{2}$ (decimal answer)
(35, 43)

19. $(-4) + (-1) - (-7) - (+8)$
(68)

20. $2\frac{1}{4} \div \left(1\frac{1}{2} \cdot 3\right)$
(26)

15

Also take Facts Practice Test O
(Classifying Quadrilaterals and Triangles).

Name _____

1. Write a proportion to solve this problem: A piece of equipment that weighs 300 pounds on Earth would
(72) weigh 50 pounds on the Moon. If an astronaut weighs 180 pounds on Earth, what would the astronaut
weigh on the Moon?

2. What is the average of the 2 numbers marked by arrows on this number line?
(28, 34)

Refer to the trapezoid at right for problems 3 and 4.

3. What is the perimeter of the trapezoid?
(19)

4. What is the area of the trapezoid?
(75)

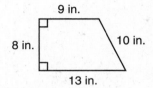

5. Four squared is how much greater than the square root of 4?
(20)

6. Use a ratio box to solve this problem: Five hundred fifty children attended the concert. If the ratio of boys
(65) to girls at the concert was 4 to 7, how many girls attended the concert?

7. Use a unit multiplier to convert 8.3 g to milligrams.
(50)

8. Diagram this statement. Then answer the questions that follow.
(71)

In the first third of the season the Wildcats played 12 games.

(a) How many games did the Wildcats play during the whole season?

(b) If the Wildcats won 75% of their games during the whole season, how many games did they win?

Write and solve equations for problems 9 and 10.

9. Thirty-six is four fifths of what number?
(74)

10. One tenth of what number is 291?
(74)

11. Simplify:
(73)

(a) $-8(-5)$

(b) $-7(+2)$

(c) $\dfrac{-15}{-5}$

(d) $\dfrac{18}{-3}$

12. If each edge of a cube is 9 cm, what is the volume of the cube?
(70)

13. Find the circumference of each of these circles.
(66)

(a)

52 in.

Use 3.14 for π.

(b)

26 m

Leave π as π.

14. Complete this table.
(48)

Fraction	Decimal	Percent
$\frac{5}{6}$	(a)	(b)
(c)	0.65	(d)

15. Evaluate: $10m - (my - y^2)$ if $m = 11$ and $y = 8$
(52)

16. Solve: $\dfrac{3}{4}y = 24$
(Inv. 7)

17. Solve: $s + 1.6 = 5$
(Inv. 7)

Simplify:

18. $5\dfrac{1}{9} \div \left(3\dfrac{1}{2} + 4\dfrac{1}{6}\right)$
(26, 30)

19. $(-6) - (-8) + (-7)$
(68)

20. $\dfrac{\$524}{1 \text{ wk}} \cdot \dfrac{1 \text{ wk}}{5 \text{ days}} \cdot \dfrac{1 \text{ day}}{8 \text{ hr}}$
(53)

16

Also take Facts Practice Test P
(+ − × ÷ Integers).

Name _____

1. Camille mowed lawns for 3 hours and earned \$6.80 per hour. Then she washed windows for 2 hours and earned \$6.20 per hour. What were Camille's average earnings per hour for all 5 hours?
(55)

2. Evaluate: $x + (x^2 − xy) − y$ if $x = 8$ and $y = 5$
(52)

3. Compare: $a \bigcirc b$ if $a − b = 0$
(79)

Use ratio boxes to solve problems 4 and 5.

4. When Shawn cleaned his room he found that the ratio of clean clothes to dirty clothes was 3 to 5. If 48 articles of clothing were discovered, how many were clean?
(65)

5. In 25 minutes, 600 customers entered the attraction. At this rate, how many customers would enter the attraction in 1 hour?
(72)

6. The diameter of a round skating rink is 18 m. Find the circumference of the rink to the nearest meter.
(66)

7. The vertices of $\triangle ABC$ are $A(2, −1)$, $B(2, 4)$, and $C(5, 4)$. Draw the triangle and its image, $\triangle A'B'C'$, reflected in the y-axis. What are the coordinates of the vertices of $\triangle A'B'C'$?
(80)

8. Graph $x < −3$ on a number line.
(78)

9. Melanie needed 28 inches of wire for her project. She used $\frac{1}{4}$ of a full spool of wire. How many inches of wire were on the full spool?
(71)

10. Simplify:
(73)

(a) $\dfrac{350}{−7}$

(b) $\dfrac{−880}{−11}$

(c) $14(−40)$

(d) $19(+60)$

11. Complete this table.
(48)

Fraction	Decimal	Percent
$\frac{4}{9}$	(a)	(b)
(c)	0.85	(d)

12. Find the area of this trapezoid. Dimensions are in meters.
(75)

Write and solve equations for problems 13 and 14.

13. Four hundred twenty is $\dfrac{3}{5}$ of what number?
(74)

14. What percent of 60 is 45?
(77)

15. Solve: $\dfrac{4}{5}m = 52$
(Inv. 7)

16. Solve: $3.5 = x − 0.09$
(Inv. 7)

Simplify:

17. $\dfrac{6\frac{1}{4}}{100}$
(76)

18. $\dfrac{3^3 + 4 \cdot 2 − 5 \cdot 2^2}{\sqrt{3^2 + 4^2}}$
(52)

19. $6\frac{1}{3} \div 1.9$ (fraction answer)
(26, 43)

20. $−33 − (−23) + (+32)$
(68)

1. The team's ratio of games won to games played was 2 to 7. If the team played 49 games, how many
(65) games did the team fail to win?

2. Find the (a) mean, (b) median, (c) mode, and (d) range of the following scores:
(28, Inv. 4)
$$80, 90, 80, 65, 95, 90, 75, 100, 65, 70$$

3. Jenny was chagrined to find that the ratio of dandelions to peonies in the garden was 12 to 5. If there were
(54) 45 peonies in the garden, how many dandelions were there?

4. Use a unit multiplier to convert 0.37 liters to milliliters.
(50)

5. Graph $x \geq -2$ on a number line.   **6.** Collect like terms: $6xy - xy - 5x + x$
(78)   (84)

Use ratio boxes to solve problems 7 and 8.

7. If sound travels 2 miles in 10 seconds, how far does sound travel in 4 minutes?
(72)

8. At the ball game, 65% of the fans waved pom-poms. If 133 fans did not wave pom-poms, how many fans
(81) were there in all?

9. Diagram this statement. Then answer the questions that follow.
(71)
> *Eighty-eight thousand dollars was raised in the charity drive. This was four fifths of the goal.*

(a) The goal of the charity drive was to raise how much money?

(b) The drive fell short of the goal by what percent?

10. A certain rectangular prism is 9 inches long, 4 inches wide, and 2 inches high. Sketch the figure and find
(70) its volume.

11. Find the area of this circle.   **12.** Complete this table.
(82)   (48)

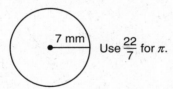
Use $\frac{22}{7}$ for π.

FRACTION	DECIMAL	PERCENT
$\frac{11}{20}$	(a)	(b)
(c)	(d)	7%

13. Multiply and write the product in scientific notation: $(2.2 \times 10^5)(2.1 \times 10^6)$
(83)

Refer to the figure at right for problems 14 and 15.

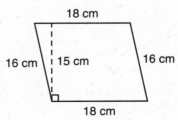

14. Name this quadrilateral and find its perimeter.
(19, Inv. 6)

15. Find the area of this quadrilateral.
(61)

Solve:

16. $18.2 = 1.4p$ **17.** $z + \dfrac{5}{8} = 1\dfrac{1}{4}$
(Inv. 7)   (Inv. 7)

Simplify:

18. $2\{45 - [6^2 - 3(11 - 7)]\}$ **19.** $2.3 \div \left(6\dfrac{1}{2} - 3\dfrac{5}{8}\right)$ (fraction answer)
(63)   (43)

20. $(-5) + (-9) - (-6) + (-1)$
(64)

TEST

18

Also take Facts Practice Test R (Area).

Name _____

1.
(54)
In the forest there were lions and tigers and bears. The ratio of lions to tigers was 4 to 5. The ratio of tigers to bears was 3 to 5. If there were 12 lions, how many bears were there? (*Hint:* First find how many tigers there were.)

2.
(70)
The shoe box was 38 cm long, 20 cm wide, and 14 cm high. What was the volume of the shoe box?

3.
(44)
A baseball player's batting average is found by dividing the number of hits by the number of at-bats and rounding the result to the nearest thousandth. If Sondra had 28 hits in 76 at-bats, what was her batting average?

4.
(88)
Use two unit multipliers to convert 27 square feet to square yards.

5.
(86)
Graph the negative integers greater than –5.

6.
(17)
How many degrees is $\frac{1}{5}$ of a full circle?

7.
(71)
Diagram this statement. Then answer the questions that follow.

Don bought the bookshelf for $42. This was $\frac{3}{4}$ of the regular price.

(a) What was the regular price of the bookshelf?

(b) Don bought the bookshelf for what percent of the regular price?

8.
(40)
Use the information in the figure at right to answer questions (a) and (b).

(a) What is m∠*w*?

(b) What is m∠*z*?

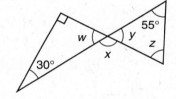

9.
(89)
How many diagonals can be drawn from one vertex of a regular hexagon? Illustrate your answer.

10.
(66)
What is the circumference of this circle?

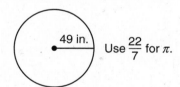

Use $\frac{22}{7}$ for π.

11.
(75)
Find the area of this trapezoid.

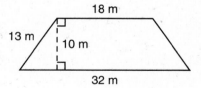

12.
(79)
Compare: $a^2 \bigcirc a$ if $a = 0.7$

13.
(48)
Complete this table.

Fraction	Decimal	Percent
(a)	0.05	(b)

14.
(77)
What percent of 200 is 50?

15.
(81)
Use a ratio box to solve this problem: Twenty-five percent of the 5000 fast-food customers ordered a hamburger. How many of the customers did not order a hamburger?

16.
(83)
Multiply and write the product in scientific notation: $(2.75 \times 10^{-2})(6 \times 10^{-8})$

17.
(Inv. 7)
Solve: $7\frac{1}{2}y = 75$

18.
(Inv. 7)
Solve: $14.2 = 6.84 + f$

19.
(87)
Simplify: $(-8x^2)(-2xy^2)$

20.
(85)
Simplify: $(-6) - (-8)(+4) - (-2)(-5)$

Saxon Math 8/7—Homeschool

19

*Also take Facts Practice Test S
(Scientific Notation).*

Name _____

1. After 4 tests Jerome's average score was 83. What score must he earn on his next test to have a 5-test
(55) average of 85?

2. Seventy-five of the 90 children in the club were girls. What was the ratio of boys to girls in the club?
(36, 65)

3. Four dozen juice bars cost $5.60. At that rate what would be the cost of 72 juice bars?
(53)

4. Maria will flip a coin three times. What is the probability that the coin will land heads up all three times?
(94)

5. Because of the unexpected cold weather, the cost of lettuce increased 25 percent in one month. If the cost
(92) after the increase was 80¢ per pound, what was the cost before the increase?

6. Write an equation to solve this problem: Thirty is what percent of 40?
(77)

7. Use two unit multipliers to convert 2000 cm^2 to mm^2.
(88)

8. If $x = -7$ and $y = 3x - 2$, then y equals what number?
(85, 91)

9. Find the volume of this prism. Dimensions are in
(95) centimeters.

10. Complete this table.
(48)

FRACTION	DECIMAL	PERCENT
$3\frac{3}{4}$	(a)	(b)
(c)	(d)	$5\frac{1}{2}\%$

11. The price of the ski jacket was $85.00. The tax rate was 6%.
(46)
  (a) What was the tax on the ski jacket?
  (b) What was the total price of the ski jacket including tax?

12. Multiply and write the product in scientific notation: $(2 \times 10^{-4})(9 \times 10^7)$
(83)

13. Graph the whole numbers less than 2.
(86)

Solve:

14. $4\frac{2}{3}x = 56$
(90)

15. $3m - 37 = 47$
(93)

Simplify:

16. $(4 \cdot 3)^2 - 4(3)^2$
(63)

17. $(-3x^2)(3x^2y)(-2xy)$
(87)

18. $4 - \left(7\frac{1}{2} - 5.6\right)$ (fraction answer)
(30, 43)

19. $2x - y + x - y$
(84)

20. $\dfrac{2 - 4 + 1 - 14 + 7(-6)}{3}$
(91)

1. Aaron's average score on the first 4 tests was 86. His average score on the next 2 tests was 92. What was
(55) Aaron's average score on all 6 tests?

2. Use a ratio box to solve this problem: After working 6 months, Martha received a raise of 20%. If
(92) Martha's previous pay was $6.90 per hour, what was her hourly pay after the raise?

3. Write an equation to solve this problem: Twenty-one is what percent of 14?
(77)

4. Use two unit multipliers to convert 2.6 m^2 to cm^2.
(88)

5. Diagram this statement. Then answer the questions that follow.
(71)
> *Seven eggs were cracked. This was $\frac{1}{6}$ of the total number of eggs in the flat.*

(a) How many eggs were in the flat?

(b) What percent of the eggs in the flat were not cracked?

6. Evaluate: $\dfrac{a + b}{c}$ if $a = -5$, $b = -4$, and $c = -6$
(91)

7. The perimeter of a certain square is 36 inches. Find the area of the square in square inches.
(19, 20)

8. The face of this spinner is divided into sixths. If the
(94) spinner is spun twice, what is the probability that the
arrow will stop on a consonant both times?

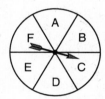

9. Find the volume of this triangular prism.
(95) Dimensions are in centimeters.

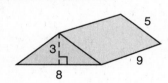

10. Find the area of this circle.
(82)

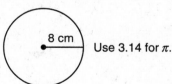

Use 3.14 for π.

11. Find the total cost, including 7% tax, of 20 square yards of carpeting priced at $17.00 per square yard.
(46)

12. What is $33\frac{1}{3}\%$ of $33.00?
(60)

13. At 5:00 a.m. the hands of a clock form an angle that measures how many degrees?
(7)

14. Multiply and write the product in scientific notation: $(3 \times 10^4)(6 \times 10^{-7})$
(83)

15. Solve: $0.5j - 1.2 = 1.2$
(93)

16. Solve: $\dfrac{2}{3}x - 3 = 15$
(93)

17. Simplify: $4^3 - \sqrt{25} + 9 \cdot 2^4$
(52)

18. Simplify: 5 yd 1 ft 11 in. + 6 in.
(49)

19. Simplify: $3x + 8(x + 2)$
(96)

20. Simplify: $\dfrac{-4(-6) + 5(-1)(-3)}{(-3)}$
(85)

1. The dinner bill totaled $21.00. Beth left a 15% tip. How much money did Beth leave for a tip?
(46)

2. The 245-kilometer drive took $3\frac{1}{2}$ hours. What was the average speed of the drive in kilometers per hour?
(46, 76)

Use ratio boxes to solve problems 3–5.

3. The $\frac{1}{48}$-scale model of the building stood 11 inches high. What was the height of the actual building?
(98)

4. Rachel saved $45 buying the dress at a 20%-off sale. What was the regular price of the dress?
(92)

5. A merchant bought an item for $30.00 and sold it for 60% more. For what price did the merchant sell the item?
(92)

6. What is 6.5% of $74.00?
(60)

7. Find the perimeter of this figure. Use 3.14 for π.
(104) Dimensions are in centimeters.

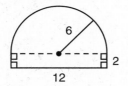

8. Use the Pythagorean theorem to find a.
(99) Dimensions are in inches.

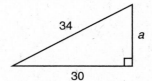

9. Find the surface area of this prism. Dimensions
(105) are in inches.

10. Find the volume of this cylinder. Use 3.14 for π.
(95) Dimensions are in centimeters.

11. These two triangles are similar. Find x.
(97)

12. Find the measure of $\angle AOB$ in this figure.
(101)

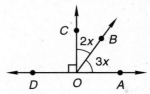

13. (a) Arrange in order from least to greatest: $5, 5^2, \sqrt{5}, -5$
(100)
   (b) Which of the numbers in part (a) is irrational?

14. If Juan flips a fair coin twice, what is the chance that the coin will land tails up twice?
(Inv. 10)

15. Which of these numbers is between 6 and 8?
(100)
   A. $\sqrt{7}$ B. $\sqrt[3]{125}$ C. $\sqrt{57}$

Solve:

16. $4x - 16 + x = 24$
(102)

17. $\dfrac{16}{w} = \dfrac{94}{4.7}$
(39)

Simplify:

18. $\dfrac{(4x^2 y)(3x^2)}{6x^2}$
(87, 103)

19. $(-3)^2 - 2^3$
(20, 103)

20. $\dfrac{(-18) - (-4)(+5)}{(-4) - (+5) - (+5)}$
(85)

22

Name _____

1. Find the (a) mean, (b) median, (c) mode, and (d) range for the following temperatures:
(28, Inv. 4)

$$60, 46, 64, 69, 72, 66, 72, 59, 68$$

2. If two cards are drawn from a normal deck of 52 cards and not replaced, what is the probability that both
(94) cards will be clubs?

Use ratio boxes to solve problems 3–5.

3. Simone can exchange $160 for 300 Swiss francs. At that rate, how many dollars would a 330-franc Swiss
(54) watch cost?

4. The pond was teeming with ducks and geese. The ratio of ducks to geese was 4 to 9. If there were 351 ducks
(65) and geese in the pond, how many ducks were there?

5. During the off-season, the room rates at the resort were reduced 30%. If the usual rate was $130 per day,
(92) what was the off-season rate?

6. Find the volume of this right circular cylinder.
(95) Use 3.14 for π. Dimensions are in centimeters.

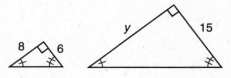

7. Use the formula $t = 1.04p$ to find t when p is 7.5.
(108)

8. Make a table that shows 3 pairs of numbers for the function $y = 2x - 3$. Then graph these pairs on a
(Inv. 9, 107) coordinate plane, and draw a line through these points. What is the slope of the graphed line?

9. What is 6.5% of $70.00?
(60)

10. Ten percent of what number is 220?
(77)

11. Find the perimeter of this figure. Use 3.14 for π.
(104) Dimensions are in centimeters.

12. In this figure lines l and m are parallel. If
(102) m$\angle a$ is 117°, then what is m$\angle h$?

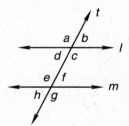

13. Solve the equation $d = rt$ for t.
(106)

14. Find m$\angle x$ in this figure.
(40)

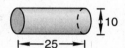

15. The triangles are similar. (a) Find y. (b) Find
(97, 98) the scale factor from the smaller triangle to the larger triangle.

16. Solve: $1\dfrac{3}{7}x - 23 = 27$
(93)

17. Solve: $4x + 10 = 2x - 14$
(102)

Simplify:

18. $\dfrac{(-7) - (8)(-3) - (-1)^2}{(-1) + (-3)}$
(85)

19. $100 - \{70 - 2[3 + 2(3^2)]\}$
(63)

20. $\dfrac{(-6de)(15d^2e)}{-15d^3e}$
(87, 103)

23

Also take Facts Practice Test W (Multiplying and Dividing in Scientific Notation).

Name _____

Use ratio boxes to solve problems 1–3.

1. The regular price was $45.00, but the skateboard was on sale for 20% off. What was the sale price?
(92)

2. If 20 kilograms of seed cost $31, how much would 30 kilograms cost at the same rate?
(54)

3. A laser printer was on sale for 40% off the regular price. If the sale price was $384, what was the regular price?
(92)

4. Divide 9×10^7 by 3×10^4 and write the quotient in scientific notation.
(111)

5. The median of these numbers is how much less than the mean?
(28, Inv. 4)
$$1.5, 0.4, 0.8, 0.85, 3.4$$

6. What is the probability of having a coin turn up heads on four consecutive tosses?
(94)

7. Teesha left $3500 in an account that paid 6% interest compounded annually. How much interest did Teesha earn in 2 years?
(110)

8. What percent of $25 is $7.75?
(77)

9. An aquarium with the dimensions shown is filled with water.
(115)
(a) How many liters of water are in the aquarium?
(b) How many kilograms of water are in the aquarium?

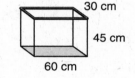
30 cm
45 cm
60 cm

10. Use two unit multipliers to convert 5 ft^2 to square inches.
(88)

11. If Jo walks from point *A* to point *B* to point *C*, she walks 84 yards. How many yards would Jo save by taking the shortcut from *A* to *C*?
(112)

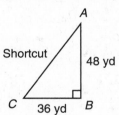
A
Shortcut
48 yd
C
36 yd
B

12. Find the volume of this pyramid. The square base is 20 m by 20 m. The height is 15 m.
(113)

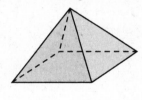

13. Make a table that shows 3 pairs of numbers for the function $y = -x - 2$. Then graph the number pairs on a coordinate plane, and draw a line through the points to show other number pairs of the function. What is the slope of the graphed line?
Inv. 9, 107

14. Use the formula $A = \frac{1}{2}bh$ to find *h* when $A = 30$ and $b = 6$.
(108)

15. Find m∠*x*.
(40)

16. Solve: $1\frac{2}{5}w - 18 = 31$
(90, 93)

17. Solve and graph on a number line: $2x + 4 \leq 6$
(93)

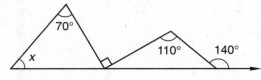
70°
x
110°
140°

Simplify:

18. $(-2)^2 \cdot 3^{-2}$
(52, 103)

19. $\dfrac{4x \cdot 4x}{4x + 4x}$
(87, 103)

20. $\dfrac{(-4) + (-7) + (3)(-3)}{(-6) - (-2)}$
(85)

Recording Forms

The five optional recording forms in this section may be photocopied to provide the quantities needed by you and your student.

Recording Form A: Facts Practice

This form helps your student track his or her performances on Facts Practice Tests throughout the year.

Recording Form B: Lesson Worksheet

This single-sided form is designed to be used with daily lessons. It contains a checklist of the daily lesson routine as well as answer blanks for the Warm-Up and Lesson Practice.

Recording Form C: Mixed Practice Solutions

This double-sided form provides a framework for your student to show his or her work on the Mixed Practices. It has a grid background and partitions for recording the solutions to thirty problems.

Recording Form D: Scorecard

This form is designed to help you and your student track scores on daily assignments and cumulative tests.

Recording Form E: Test Solutions

This double-sided form provides a framework for your student to show his or her work on the tests. It has a grid background and partitions for recording the solutions to twenty problems.

A Facts Practice

Name _____

TEST	# POSSIBLE	TIME AND SCORE (time / # correct)								
A 64 Multiplication Facts	64									
B 30 Equations	30	B	B	B	B	C	C	C	C	C
C 30 Improper Fractions and Mixed Numbers	30									
D 40 Fractions to Reduce	40	D	D	D	D	D	E	E	E	E	E
E Circles	24									
F Lines, Angles, Polygons	16									
G + − × ÷ Fractions	24									
H Measurement Facts	33	H	H	H	H	H	I	I	I	I	I
I Proportions	24									
J + − × ÷ Decimals	21	J	J	J	J	J	K	K	K	K
K Powers and Roots	32									
L Fraction-Decimal-Percent Equivalents	25	L	L	L	L	L	M	M	M	M	M
M Metric Conversions	24									
N + − × ÷ Mixed Numbers	20									
O Classifying Quadrilaterals and Triangles	9	O	O	O	O	O	P	P	P	P	P
P + − × ÷ Integers	32									
Q Percent-Decimal-Fraction Equivalents	25	Q	Q	Q	Q	Q	Q	R	R	R	R
R Area	12									
S Scientific Notation	20	S	S	S	S	T	T	T	T	T
T Order of Operations	16									
U Two-Step Equations	15									
V + − × ÷ Algebraic Terms	18									
W Multiplying and Dividing in Scientific Notation	18									

B

Lesson Worksheet
Show all necessary work. Please be neat.

Name _____

Date _____

Lesson _____

Warm-Up
- ☐ Facts Practice
- ☐ Mental Math
- ☐ Problem Solving

Review
- ☐ Homework Check
- ☐ Error Correction

Instruction
- ☐ Lesson
- ☐ Lesson Practice
- ☐ Mixed Practice

Facts Practice

Test:	Time:	Score:

Mental Math

a.	b.	c.	d.	e.	f.
g.	h.	i.	j.	k.	l.

Problem Solving

Strategies:
(Check any you use.)

- ☐ Make a chart, graph, or list.
- ☐ Guess and check (trial and error).
- ☐ Use logical reasoning.
- ☐ Act it out. ☐ Draw a diagram.
- ☐ Make it simpler. ☐ Draw a picture.
- ☐ Work backward. ☐ Find a pattern.

Lesson Practice

a.	b.	c.
d.	e.	f.
g.	h.	i.
j.	k.	l.

Saxon Math 8/7—Homeschool

C | Mixed Practice Solutions
Show all necessary work. Please be neat.

Name _____

Date _____

Lesson _____

2.

3.

5.

6.

8.

9.

11.

12.

14.

15.

16.

17.

18.

19.

20.

21.

22.

23.

24.

25.

26.

27.

28.

29.

30.

D	**Scorecard**	Name _____

Date	Lesson or Test	Score	Date	Lesson or Test	Score	Date	Lesson or Test	Score	Date	Lesson or Test	Score

E | Test Solutions

Show your work on this paper.
Do not write on the test.

Name _____

Date _____

Test _____ Score _____

2.

4.

6.

8.

10.

11.

12.

13.

14.

15.

16.

17.

18.

19.

20.